Digital Decarbonization
Promoting Digital Innovations
to Advance Clean Energy Systems

COUNCIL *on*
FOREIGN
RELATIONS

Maurice R. Greenberg
Center for Geoeconomic Studies

June 2018

Edited by Varun Sivaram

Digital Decarbonization
Promoting Digital Innovations
to Advance Clean Energy Systems

The Council on Foreign Relations (CFR) is an independent, nonpartisan membership organization, think tank, and publisher dedicated to being a resource for its members, government officials, business executives, journalists, educators and students, civic and religious leaders, and other interested citizens in order to help them better understand the world and the foreign policy choices facing the United States and other countries. Founded in 1921, CFR carries out its mission by maintaining a diverse membership, including special programs to promote interest and develop expertise in the next generation of foreign policy leaders; convening meetings at its headquarters in New York and in Washington, DC, and other cities where senior government officials, members of Congress, global leaders, and prominent thinkers come together with CFR members to discuss and debate major international issues; supporting a Studies Program that fosters independent research, enabling CFR scholars to produce articles, reports, and books and hold roundtables that analyze foreign policy issues and make concrete policy recommendations; publishing *Foreign Affairs*, the preeminent journal of international affairs and U.S. foreign policy; sponsoring Independent Task Forces that produce reports with both findings and policy prescriptions on the most important foreign policy topics; and providing up-to-date information and analysis about world events and American foreign policy on its website, CFR.org.

The Council on Foreign Relations takes no institutional positions on policy issues and has no affiliation with the U.S. government. All views expressed in its publications and on its website are the sole responsibility of the author or authors.

For further information about CFR or this paper, please write to the Council on Foreign Relations, 58 East 68th Street, New York, NY 10065, or call Communications at 212.434.9888. Visit CFR's website, CFR.org.

This publication was made possible by a grant from the Alfred P. Sloan Foundation.

Contents

Introduction

Varun Sivaram

In 2017, the *Economist* proclaimed that data was the new oil.[1] Just as trade in oil has underpinned the global economy for a century, flows of data—the most valuable resource of the twenty-first century—now drive economic value. Indeed, in 2017, all five of the world's most valuable publicly traded companies specialized in digital technologies, whereas just a decade earlier three of the top five companies were in the energy sector.

This does not mean that the energy sector has been left behind by the digital revolution. To the contrary, digitalization is at the heart of the tectonic shifts that are starting to reshape the energy landscape. As energy industries produce ever more data, firms are harnessing greater computing power, advances in data science, and increased digital connectivity to exploit that data. These trends have the potential to transform the way energy is produced, transported, and consumed.[2]

An important potential benefit of this digital transformation of energy is a reduction in global emissions of greenhouse gases that cause climate change; the elimination of such emissions from the global economy is known as decarbonization. By enabling clean energy systems that rely on low-carbon energy sources and are highly efficient in using energy, digital innovations in the energy sector can speed decarbonization. Yet they are not guaranteed to do so. In fact, digital innovations could well increase global greenhouse emissions, for example, by making it easier to extract fossil fuels.[3]

To determine the potential for digital technologies to speed a clean energy transition and to make recommendations to promote this outcome, the Council on Foreign Relations convened a workshop in New York, on February 22 and 23, 2018. The gathering included nearly forty current and former government officials, entrepreneurs, scientists, investors, executives, philanthropists, and policy researchers from around the world. Participants laid out a wide range of areas in which digital technologies are already enabling clean energy systems and could achieve much more; they also cautioned about serious risks that will attend the digitalization of energy and need to be managed; and they articulated actionable recommendations for policymakers in the United States and abroad to ensure that digital innovations bring societal benefits and, in particular, speed decarbonization.

To guide the discussion, participants produced twelve essays on various topics related to digitalization and clean energy systems. This volume compiles those essays into a narrative spanning opportunities, challenges, and recommendations.

THE DIGITAL WAVE
OF CLEAN ENERGY INNOVATION

Investment in clean energy innovation has been volatile over the last decade. In 2008, venture capital investment in clean energy start-up companies reached record levels as private investors sought to transform global energy systems by funding innovative technologies to produce solar power, store energy in batteries, harness biofuels to displace petroleum, and more. But then their investments failed at a staggering rate: venture capital investors lost more than half of the $25 billion they had invested in clean energy technology companies between 2006 and 2011. As a result, they sharply pulled back funding for the sector.[4]

Yet as Stephen D. Comello argues in his essay, a new crop of innovative clean energy start-ups in the electric power sector is rising from the ashes of the failed investments from the last decade. His data shows that a range of investors—from venture capitalists to electric power utilities—are ramping up funding for a new wave of clean energy innovation. Comello argues that investments in technologies to produce renewable energy (such as new solar energy materials), which he dubs wave 1 investments, are substantively distinct from today's wave 2 investments, which might actually make money for investors.

Whereas wave 1 was about energy supply, wave 2 is all about energy delivery and consumption. Comello argues that digital technologies—including digital communication, data analytics, and system automation software—are making it possible to operate electric grids more efficiently and enable consumers to use less energy. And in contrast to wave 1 companies, which required massive amounts of investment capital to scale up new energy materials and manufacturing processes, wave

2 companies have so far required far less capital—less than half, according to Comello's data. Therefore, investors can afford to spread their capital over a broader portfolio of companies to hedge their risk. They might even realize investment returns more quickly because a wave 2 company might be able to bring a digital innovation to market faster than a wave 1 company could commercialize a breakthrough in laboratory science.

Most workshop participants were cautiously optimistic about this digital wave of clean energy innovation, but some raised concerns about its viability. One pointed out that firms in the energy sector, such as electric power utilities, are generating tremendous volumes of data, but often that data is not produced in standard formats across or even within firms. Improvements in the quality and standardization of data will be needed for start-ups to meaningfully harness that data. Another participant argued that collaboration between electricity utilities and start-ups, while essential for digital innovations to be adopted at scale, is complicated by cultural differences. Utilities seek to ensure reliable electric service to customers and therefore prize stability, whereas start-ups are focused on rapid change. Working together will require that each side understand the other's priorities.

David Victor's essay also raises concerns that wave 2 start-ups might not succeed in overhauling energy systems as rapidly as optimists estimate. He argues that "Silicon Valley is better at silicon than at crossing valleys," asserting that successful firms that have emerged from the Silicon Valley innovation model—from Intel to Facebook—have created new markets and established natural monopolies. In energy, by contrast, start-ups seek to conquer markets that have already been monopolized by incumbent firms such as electric power utilities. Doing so is a tall order, and Victor advises reserving judgment on the success of the new wave 2 investment thesis until the actual performance of start-ups can be evaluated.

Victor's most pointed rebuke is that digitalization is not necessarily going to benefit clean energy systems any more than dirty energy ones. He predicts that digital innovations will have far-reaching and unpredictable effects, but he also argues that digitalization is an "equal opportunity revolution." For example, firms are using digital technologies to reduce the cost of extracting oil and gas, which could enable greater consumption of fossil fuels and overwhelm the carbon savings from other digital innovations that increase energy efficiency. Victor concludes that digitalization can decisively favor clean energy systems only if public policies incentivize decarbonization. Unfortunately, he laments, the increasingly powerful renewable energy lobby does not advocate for such policies, preferring instead to target narrow incentives for renewable energy that would not promote a range of innovations, including digital ones, to cost-effectively slash carbon emissions.[5]

Workshop participants had a spirited discussion about whether digital innovations would naturally advance decarbonization without public policies such as a carbon price. Despite broad agreement that digitalization could help fossil fuel companies, some participants postulated reasons why the effects of digital innovations might skew toward reducing carbon emissions. One suggested that digital technologies promote efficiency, and using less energy—clean or dirty—will reduce carbon emissions. Another added that digitalization enables more decentralized energy systems—such as electric power microgrids—and that such systems can be more efficient in their energy consumption and also rely on clean energy sources such as distributed solar panels. Not everyone agreed, however. In a poll of the experts at the workshop, 27 percent dissented from the thesis that "digital innovations are very likely to advance, rather than hinder, clean energy and decarbonization" (see figure 1).

FIGURE 1. EXPERTS' ASSESSMENT OF THE STATEMENT: "DIGITAL INNOVATIONS ARE LIKELY TO ADVANCE, RATHER THAN HINDER, CLEAN ENERGY AND DECARBONIZATION"

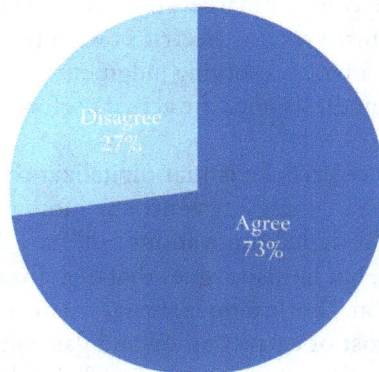

Source: Survey of workshop participants.

DIGITAL OPPORTUNITIES

Although digital innovations cannot be guaranteed to advance clean energy systems and reduce emissions, opportunities for them to do so are abundant. Workshop participants focused on the opportunities for digital innovations to decarbonize electric power and transportation systems.

ELECTRIC POWER

Participants were most enthusiastic about the opportunities for digital innovations to transform the electric power system and make it more efficient, resilient, and clean. Some of the enormous value of these opportunities arises simply because electricity systems around the world are both massive and antiquated. Basic digital upgrades, such as equipping power plants and the grid with sensors to enable more efficient operation, could reduce the amount of wasted energy in power generation and delivery as well as cut system costs. Already, in 2016, global investment in digital electricity infrastructure and software ballooned to $47 billion, a figure greater than the amount spent on natural gas-fired power plants.[6]

But, as Lidija Sekaric writes in her essay—a survey of digital opportunities in the power sector—upgrading the existing electricity system is only the first step in an electric power revolution. The most important effect of digital innovations, she asserts, will be the decentralization of power systems around the world. She argues that "a fully transactive grid of the future could empower prosumers (consumers of energy who also produce it) to trade electricity at the edges of the grid." That future would stand in stark contrast to the existing model of centralized generation of electricity, followed by one-way transmission and distribution to customers.

Sekaric lays out an incremental path for digital innovations to gradually decentralize the grid. She foresees digital technologies in the coming years enabling the centralized grid to integrate the proliferation of distributed energy resources, such as solar panels. She also argues that electric system operators will be able to marshal multitudes of internet-connected electric appliances and battery banks, which can modulate their energy use (and output, in the case of batteries) in aggregate to act as a virtual power plant. Such plants could rapidly compensate for the

intermittent output of wind and solar power, enabling a greater penetration of such clean energy sources. Finally, Sekaric is bullish about
microgrids, which use digital technologies to balance demand with
supply from local energy sources. As a bonus, advanced microgrids can
seamlessly disconnect from the main grid—for example, to keep critical facilities running in the face of a disaster—and to provide support to
the grid during connected operation. These advances are held back, she
argues, by power sector regulations that have been slow to keep up with
the pace of technological improvement and business model innovation.

Ben Hertz-Shargel elaborates in his essay on the conditions needed
to enable such a decentralized electric power system. He argues that the
current rise of distributed energy resources threatens the stability of
the grid by making it difficult to balance supply and demand, especially
at local scales. For example, electric vehicles can increase demand for
power beyond what a distribution circuit can deliver; conversely, rooftop solar panels can send power from customers back to the grid, over-
loading equipment intended to only handle power flowing in a single
direction. But Hertz-Shargel also recognizes that distributed energy
resources can be orchestrated to restore stability to the grid.

He argues that new economic markets—known as distribution
markets—are crucial to realizing a future in which customers deploy
their distributed energy resources to help balance supply and demand
at all points in the electricity network. A system operator would monitor real-time operational data across the distribution grid and calculate
prices for electricity services every few minutes at each local node in the
network. Customers would then transact in those markets; as Hertz-
Shargel foresees it, smart software agents would save customers the
trouble of managing their energy use and distributed energy resources
themselves, by instead controlling multiple customer devices for them
simultaneously to most efficiently clear the market and save customers
money. Ultimately, dynamic distribution markets could drive further
innovation in the digital technologies needed to manage distributed
energy resources. In turn, utilities would have less need to maintain an
expensive, centralized electricity network.

During the workshop, some participants called this futuristic vision
of a decentralized, highly efficient, and transactive electricity system
a utopia. Participants stressed the serious obstacles to achieving this
vision, including the need to deploy substantial infrastructure such as
smart electricity meters and grid sensors as well as the technical and

political complexity of implementing distribution markets. Some questioned whether such a fully decentralized electric power system will in fact be necessary to decarbonize the electric power system. Indeed, one participant noted that plenty of low-hanging fruits—such as improving the energy efficiency of buildings—could substantially reduce emissions even without a transformation of the electric power system. Nevertheless, participants broadly agreed that in the coming years, distributed energy resources will play an increasingly important role in the historically centralized power sector.

TRANSPORTATION

The transportation sector poses an even greater decarbonization challenge than the electric power sector. Whereas cost-competitive clean power sources, such as wind and solar power, are beginning to make a dent in the global dominance of fossil-fueled power plants, oil retains a vice grip on the transportation sector. Recognizing this, the Barack Obama administration's 2016 blueprint for slashing U.S. greenhouse gas emissions foresaw eliminating electric power sector emissions almost entirely, leaving room for transportation emissions to fall less rapidly and dominate U.S. emissions by midcentury.[7]

As Peter Fox-Penner points out in his essay, encouraging the adoption of electric vehicles—thereby linking electric power and transportation—is the most important element of decarbonizing the transportation sector. Not only are electric vehicles more efficient than conventional vehicles fueled by petroleum, but if the electricity system can decarbonize rapidly, then electric vehicles will also get even cleaner. Yet he cautions that digital innovations will be needed to integrate a rapidly rising share of electric vehicles, which will otherwise strain the grid. He urges policymakers to guide investment in a network of electric charging stations that are "intelligent and well integrated with the rest of the power system," so that electric vehicles "could represent sources of electricity storage and load management as well as regulation of grid frequency and voltage to ensure [grid] reliability."

Fox-Penner notes, however, that even as the rise of electric vehicles advances decarbonization, the concurrent rise of autonomous vehicles adds substantial uncertainty about future emissions from the transportation sector. Firms including Uber, Alphabet, Tesla, and General Motors are busily harnessing digital advances in sensors, data processing, and artificial intelligence to improve autonomous technology. The effect of autonomous vehicles on transportation sector emissions is highly uncertain, Fox-Penner notes. By making travel much cheaper and more convenient, autonomy could increase the number of vehicle miles traveled. But autonomous vehicles might also encourage carpooling and help ferry passengers to clean mass transit options. Fox-Penner concludes that policymakers need to learn much more about the effects of autonomous vehicles and the cost of the infrastructure they will need, and proactively guide autonomous vehicles to complement, rather than compete with, mass transit.

Rohit T. Aggarwala drills into the potential effects of autonomous vehicles. Not everything is uncertain, he argues, and he lays out a set of expectations. For example, he is reasonably confident that taxi fleets of autonomous vehicles will emerge in cities within the next five years, followed shortly thereafter by commercial sales of autonomous vehicles. Autonomous vehicles will likely improve safety on roads, not reduce it. They will compete with public transit on low-volume routes or off-peak hours but will not replace high-density routes. Finally, autonomous vehicles will not immediately be fully electric—the digital equipment required for autonomy, such as sensors, can overtax the battery capacity of electric vehicles—but a gradual merger between electrification and autonomy could take place as batteries, sensors, and computation all improve.

Yet Aggarwala acknowledges several areas of uncertainty. Governments and the public could be reluctant to permit autonomous vehicles on roads, and fatalities in March 2018 involving Uber and Tesla vehicles further reduce certainty over when autonomous vehicles will gain acceptance.[8] In addition, Aggarwala imagines radically new business models emerging to take advantage of autonomy, such as constantly cruising warehouses offering five-minute merchandise delivery, which could dramatically alter vehicle usage. He urges policymakers to consider investing in redesigned roads and new digital infrastructure, implementing congestion pricing, and establishing standards for safe autonomous driving.

Workshop participants homed in on the massive uncertainty for future emissions that the arrival of autonomous vehicles brings. Indeed, projections vary wildly between halving and doubling emissions, depending on how much additional driving autonomous vehicles cause, how many of them are electric, and how they interact with mass transit.[9] In addition, participants pointed out that many other digital innovations could transform transportation and, in turn, influence the sector's emissions. For example, autonomous drones could alter demand for freight; so could additive manufacturing (also known as 3D printing), which could make it easier to disseminate electronic designs to locally produce goods, reducing freight demand and upending existing models of product storage and distribution.

So whereas participants were relatively confident that, in aggregate, digital innovations would advance clean energy systems, when it came to transportation they tended to believe the opposite. More than two-thirds of participants predicted that the digitalization of mobility would increase, rather than decrease, greenhouse gas emissions through 2040 (see figure 2).

BIG DATA AND DATA SCIENCE APPLICATIONS ACROSS ENERGY SECTORS

The dramatic increase in energy-related data—produced by grid sensors, smart meters, autonomous vehicles, and other digital equipment—is creating opportunities to better understand and manage complex energy systems. In parallel, advances in fields such as machine learning have made it possible to glean insights from large datasets. Workshop

FIGURE 2. EXPERTS' ASSESSMENT OF THE STATEMENT: "DIGITALIZATION OF TRANSPORTATION WILL DECREASE GREENHOUSE GAS EMISSIONS THROUGH 2040"

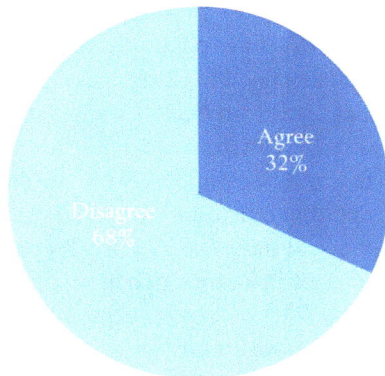

Source: Survey of workshop participants.

participants were enthusiastic about harnessing data across energy sectors to promote clean energy systems, but they cautioned that several barriers need to be surmounted in order to do so.

Kyle Bradbury lays out a taxonomy for the application of data science to energy systems in his essay. (He also makes an important distinction: big data refers to large volumes of data, whereas data science refers to the application of statistics, mathematics, and computer to extract insights from data.) He first identifies opportunities across the electric power sector. To better integrate clean electricity from intermittent wind and solar, data science can enable better forecasting of renewable output and customer demand and help system operators ensure they are balanced. Data science can also aid system operators in orchestrating large numbers of distributed energy resources. And it can help residential customers save energy by disaggregating smart meter data to determine which appliances are the most intensive energy users; it can also promote energy efficiency in buildings by informing intelligent management of building heating and cooling patterns.

But the applications of data science are not limited to individual energy sectors; rather, Bradbury argues, data science can help plan and operate complex energy systems that cross multiple sectors that might traditionally have been isolated. For example, data science can make it possible for electricity system operators to recruit fleets of electric vehicles to serve as mobile batteries to support the grid. It can also help developing countries determine how best to expand electricity access to rural communities that lack it by predicting which areas are most easily served by centralized transmission lines and which are better suited for isolated microgrids.

In his essay, Sunil Garg dives deeply into a single application of data science to promote clean energy. He shares his company's experience conducting predictive analytics on fleets of wind turbines to improve turbine operation and reduce maintenance costs. For example, Garg shares an anecdote about an instance when his company analyzed sensor data to determine that a turbine was about to fail and managed to fix the problem that could have led to a much costlier malfunction. He also provides examples of cases in which his company could increase the power production from turbines by spotting turbines that were producing less power than neighboring turbines.

Garg also shares valuable insights about the difficulties that new commercial entrants face in seeking to harness data science to disrupt the business models of incumbent equipment manufacturers. For example, U.S. copyright law is unclear about whether equipment manufacturers can restrict access to the data produced by machines. So wind turbine owners and third parties might not be legally allowed to harness turbine data to conduct predictive maintenance.

Workshop participants broadly agreed that a scarcity of nonproprietary data is a serious barrier to the application of data science to energy. Several participants noted that academic research studies often use datasets that belong to a corporation and cannot easily be shared with other researchers for the purposes of replication. For example, electric power utilities often require researchers to sign nondisclosure agreements to access and analyze electricity consumption data from smart meters.

Participants suggested that academic journals should require researchers to publish the company from which they obtained their data so that others can seek to replicate their research results. In addition, participants stressed the importance of building open-source datasets that contain energy data, so that researchers and firms can train machine learning algorithms that work as well in energy as they do in other fields. Finally, to address challenges in the energy sector, one participant proposed hosting predictive modeling and analytics competitions to elicit new approaches to harnessing data science; this approach has succeeded in other fields, such as computer vision and biology, and is conducted by organizations such as Kaggle.[10]

MANAGING THE RISKS OF DIGITAL INNOVATIONS

Alongside sweeping opportunities to advance clean energy systems, digitalization also raises serious risks. For example, the proliferation of internet-connected devices on the electric power system, from grid equipment to customer appliances, might not only enable more efficient operation of the system but also present a multitude of new access points for malicious hackers seeking to steal sensitive customer data or even take down the grid. Workshop participants grappled with how policymakers could minimize these risks—spanning cyberattacks, privacy breaches, and economic displacement—while maximizing the benefits of digitalization.

In his essay, Erfan Ibrahim lays out prudent safeguards that electric power utilities should implement to guard against a range of cyber threats that could endanger the reliable supply of electricity or expose customer data. No longer can utilities count on security-by-obscurity, a relic of an era when the electric power network had minimal digital connectivity and was hard to simply control, let alone hack. Today, he warns, the power system faces an array of threats, from foreign hackers to disgruntled insiders already in possession of access codes to the network. Therefore, utilities should overhaul their business processes to ensure cybersecurity across their operations.

In addition, Ibrahim lays out a four-layer cybersecurity architecture that establishes what he calls defense-in-depth. By this, he means that even if an attacker breaches the first layer of defense, additional safeguards will contain the attack. The first layer, including elements such as password protection, provides basic assurances that only legitimate users can access the network. The next layer, which monitors the system for the signatures of malicious software, aims to prevent and contain intruders. The third layer is even more sophisticated, protecting against insider threats by dynamically adapting to identify suspicious behavior within the network. The final layer—endpoint protection—guards against attacks on the devices and equipment at the edges of the electricity network. All of these layers collectively can help utilities ensure that the grid continues to operate reliably and safeguard customer privacy from attackers even as digitalization advances.

Although cyberattacks can threaten customer privacy, they are not the only cause for privacy concerns in an increasingly digitalized electric power system. Rather, the mere fact that digitalization could enable the collection and storage of massive amounts of customer data is unnerving to privacy advocates, who fear that firms and governments that obtain the data through legitimate means could intrude into individual privacy. In her essay, Jesse Scott surveys privacy risks across countries and suggests a path forward for policymakers. Data can be extremely revealing, she notes. For example, highly granular electricity consumption data collected by utilities through smart meters can be disaggregated to discern an individual household's consumption habits down to the appliance level or to locate empty homes, and analysis of commercial building energy consumption could reveal proprietary business practices. But the same data can also be put to good use, for example, by enabling utilities to pinpoint ways for customers to reduce their energy use or identify which customers might benefit from installing distributed generation.

No one-size-fits-all policy recommendation can help navigate this trade-off between harnessing customer data and protecting privacy, Scott argues. Rather, prudent policies will depend on cultural context. Nevertheless, Scott notes, jurisdictions can learn from one another. For example, California and South Korea have pioneered privacy protections on energy data, such as that from smart meters, and the European Union has enacted a comprehensive privacy regulation, known as the General Data Protection Regulation, that protects customers' privacy across economic sectors including energy. Finally, Scott also discusses the risk of economic displacement arising from digitalization in the energy sector. She concludes that uncertainty is substantial over whether autonomous vehicles will spur large job losses, such as in trucking. She is confident, however, that across a range of fields from power plant operation to energy-efficient equipment installation, workers will need to be trained in new skills related to digital technologies.

Workshop participants were most concerned about cybersecurity risks. Over 90 percent expressed more concern about the risk of cyberattacks than about privacy breaches or economic displacement. Participants noted that even though decentralizing the grid might improve its resilience to natural disasters and physical attacks, the proliferation of distributed energy resources might increase cybersecurity risk by creating new access points for malicious actors. One participant argued that the digital architecture of the grid should therefore be sectionalized; that is, self-sufficient zones of the grid should be independently operated by distributed computing infrastructure rather than by a system operator using a centralized cloud computing approach to control the grid.

Participants emphasized the importance of designing cybersecurity and privacy regulations specific to the energy sector, to clarify regulatory uncertainty and avoid regulations tailored for the financial services or health-care sectors governing the energy sector. One participant remarked that the protections for the data produced by an electrical substation should look markedly different from the protections afforded to a customer's personal health data. Finally, participants noted that even with privacy safeguards in place, such as anonymization of smart meter data, personal information could still be exposed if someone with access to the anonymized data were to cross-reference that data with another dataset, such as satellite imagery, to identify the energy consumption data that corresponded to particular homes. This insight suggests that policymakers need to be even more thoughtful when designing regulations to protect customer privacy while enabling the collection of data useful to advancing clean and cost-effective energy systems.

POLICY RECOMMENDATIONS

Digital innovations represent a bright spot in the landscape of energy innovation. Recognizing this, workshop participants noted that policymakers should be judicious with their interventions. One participant cautioned against treating all barriers to digitalization as nails to be struck by the hammer of public policy. The role of policy should be to set the right incentives for free markets to spur digital innovations to promote clean energy systems.

In their essay, Richard Kauffman and John O'Leary explain how New York State is incentivizing electric power utilities to innovate digitally. Historically, regulations in New York and other states have incentivized utilities to build centralized infrastructure; doing so benefits utility shareholders, who earn a return on capital expenditures. As a result, New York's electricity grid is centralized, inefficient, and underutilized. Utilities have little incentive to modernize the grid and save customers money, given that over 90 percent of the customer rates that they collect simply defray costs, rather than benefiting shareholders.

But one of the goals of New York's Reforming the Energy Vision agenda is to transform this incentive structure. New York is rewriting its regulations, Kauffman and O'Leary note, to compensate utilities whenever they save customers money—this is known as a shared savings model, in which the utility's profitability is tied directly to lower customer bills. Utilities are also encouraged to turn to private markets for creative solutions to achieve these savings. Often, the most efficient solutions involve digital innovations. For example, in the state's Brooklyn-Queens Demand Management program, the utility Con Edison is working with distributed energy resource providers to avoid building a $1 billion substation. The utility is investing in a portfolio of approaches to reduce peak energy demand, including by installing energy storage and efficient customer equipment and by making digital upgrades so that customer demands become more responsive to the grid's needs. Overall, the program could save customers hundreds of millions of dollars.

New York is not the only jurisdiction experimenting with regulatory reforms. One participant noted that in the United Kingdom regulators have turned to a sandbox approach to pilot innovative technologies and systems. In a regulatory sandbox, limited to a particular section of an electric power system, firms can test out a concept such as a peer-to-peer electricity trading system facilitated by blockchain technology without the constraints of regulations that would otherwise prevent such trading. This approach can enable a successful demonstration project that serves as an example on how to alter regulations to foster innovation.

Singapore is another leader in its embrace of digitalization, and Hiang Kwee Ho writes about the city-state's approach to becoming a "smart nation." Singapore has ambitious goals for reducing its greenhouse gas emissions, and its government recognizes that even though it has limited land on which to build renewable energy projects, digital innovations can accelerate its sustainability efforts. Ho writes that Singapore seeks to optimize the operation of electricity production, transmission, and delivery; design mobility systems that pair autonomous vehicles with mass transit; and simulate its complex and interconnected heating, power, transportation, and industrial sectors to identify opportunities for reducing emissions.

Not only will Singapore's energy-specific measures reduce its carbon footprint, but its broader digital policies should also result in efficiency gains for its energy systems. For example, the government is investing in high-speed broadband networks, sensor networks, and research and development into digital technologies such as artificial intelligence. Recognizing that Singapore cannot meaningfully influence global emissions through its own sustainability initiatives, the government is aiming to lead regional energy cooperation through the Association of Southeast Asian Nations (ASEAN). For example, Ho suggests, Singapore hopes to deploy digitalization to manage a pan-ASEAN power grid that can integrate substantial amounts of intermittent renewable energy.

Workshop participants also discussed what types of digital infrastructure governments should fund. Participants generally agreed that some equipment, such as electric power transmission sensors known as phasor measurement units, are prudent investments for governments to ensure efficient and reliable electricity delivery. Agreement was weaker on whether other types of hardware—from digital communications infrastructure to intelligent vehicle chargers—should be funded by the public or private sectors.

Several participants argued that carbon pricing was an essential policy to ensure that digital innovations do, in fact, advance clean energy systems and reduce carbon emissions. Few disagreed that digitalization has transformative potential for the energy sector and that private activity in the sector is only increasing. But the question participants kept returning to was in which direction this digital wave of innovation would carry energy systems.

Part I: The Digital Wave of Clean Energy Innovation

Trends in Early-Stage Financing for Clean Energy Innovation

Stephen D. Comello

Following the rise, commercialization, and diffusion of renewable energy technologies, digital technologies represent the next wave of innovation in the electric power sector. Through advances in computing power, data acquisition, and networking capabilities—all of which have occurred largely outside the energy industry—digital technologies promise to make power systems around the world more resilient, flexible, cost effective, and clean.

In operational terms, digital technologies comprise three basic elements: data, analysis, and connectivity. Data refers to the creation and capture of information in a digital (electronic) form. Analysis is the transformation of data via computation into human-, device-, or machine-readable results. Connectivity is the exchange of data through telecommunication networks.

Over the past decade, digital technologies have proliferated across all industries in the wake of marked cost reductions and performance enhancements. The electricity industry has long used digital technologies—in fact, utilities and grid operators were early adopters of information and communications technologies (ICT). In recent years, however, these technologies have migrated from large proprietary systems developed by a handful of legacy companies to devices at the individual building level, often disseminated by firms large and small that are newcomers to the electricity industry. Examples of offerings aimed at these markets include smart thermostats (Nest), intelligent lighting (WeMo), home hubs (Amazon Echo), real-time energy consumption and device-level control (Newatt), and interactive software for customers to modulate their demand in response to electricity grid needs (OhmConnect). Moving digital technologies closer to the end user is bringing about greater convenience, comfort, and control—as well as less energy consumption and cost—to a large number of customers. These newly empowered businesses, homeowners, and individuals are placing increasing value on such energy services, inducing further demand for innovations.

Yet the proliferation of digital technologies at the grid's edge is also causing headaches for the utilities and system operators who increasingly have to manage a dizzying array of distributed energy resources as well as a rising share of intermittent renewable energy supply. Fortunately, other digital technologies can help them do so. Indeed, in 2016, approximately $47 billion was spent globally on infrastructure and software directed toward upgrading legacy ICT systems and digitalizing the electricity sector to operate the electricity system more flexibly, integrate more renewable energy, and better manage customer demand.[1]

Start-ups are an important source of digital innovation. They are growing in number and are being supported by a range of early-stage investors. If start-ups offering digital energy solutions can hone their business models and successfully collaborate and compete with energy incumbents, then this wave of clean energy technology investment could prove more durable than the last one, in which investors poured billions into companies from 2008 to 2011 and lost most of the money. Already, this second digital wave is showing more promise than the first wave of investment. But ensuring its success and realizing the benefits of digital innovation will require collaboration among the private, nonprofit, and public sectors.

A NEW WAVE OF INVESTMENT IN CLEAN ENERGY INNOVATION

The activity of early-stage investors is a bellwether for an industry's future. Early-stage investors provide grant, seed, and series-round capital to start-ups and include venture capital (VC) firms, corporate venture capital (CVC) divisions, government funding agencies, incubators and accelerators, and high net worth individuals. Within the clean energy sector, the focus of this investor class has shifted over the past fifteen years from energy generation technologies and hardware development (wave 1 technologies) to energy end-use efficiency and digital technologies (wave 2 technologies) (see figure 1).[2]

The digital technology offerings by wave 2 start-ups can be divided into four broad categories: energy management, data analytics,

FIGURE 1. GLOBAL TRENDS IN INVESTMENT BY EARLY-STAGE CAPITAL PROVIDERS FOR WAVE 1 AND WAVE 2 FIRMS

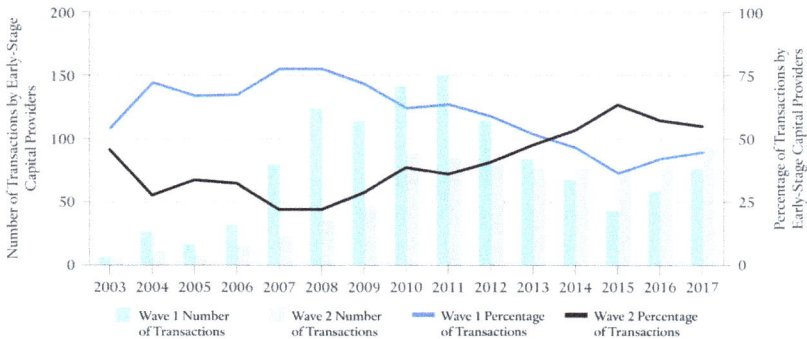

Source: Underlying data provided through CB Insights; subscription furnished by Stanford Graduate School of Business.

FIGURE 2. AVERAGE TRANSACTION SIZES FOR WAVE 1 AND WAVE 2 FIRMS

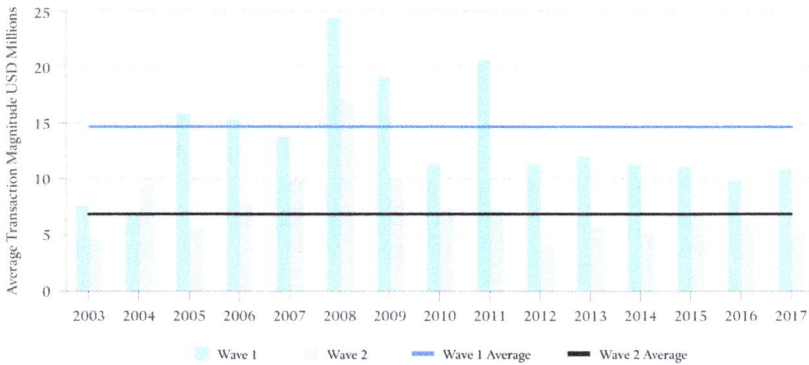

Source: Underlying data provided through CB Insights; subscription furnished by Stanford Graduate School of Business.

communication platforms, and marketplace platforms. Energy management includes sensing, data acquisition, and system control technologies across scales (home, building, cluster, system wide). Data analytics refers to the computational programs and methods used to convert data inputs into useful forms. Communication platforms refer to marketing and information-sharing forums for customers. Marketplace platforms refer to transaction and trading platforms, including distributed consensus-based platforms.

In 2017, early-stage capital providers contributed approximately $382 million in funds across ninety-four wave 2 deals. Energy management–focused start-ups received 50 percent of this capital and accounted for 43 percent of all transactions, followed by data analytics solutions providers, which accounted for 21 percent of funds received and 23 percent of transactions. The average deal size for these two categories was $4.7 million and $3.9 million, respectively, relative to $4.1 million for all digital technologies in 2017. The transaction size is instructive insofar as it represents a reduction from historical averages for both wave 1 and wave 2 clean energy technologies (see figure 2). When less capital is required to explore a business model, potentially more start-ups can be funded, ceteris paribus; this increases the probability of finding a successful venture soon. In digital technology start-ups in the energy sector, VCs are increasingly seeing the capital-light opportunities they prefer relative to the prior decade.

Of the firms that received the earliest venture funding (seed and series A) during 2016 and 2017, approximately 60 percent focused on the commercial and industrial sectors, and 25 and 15 percent primarily households and utilities, respectively. As a customer segment, commercial and industrial users are plausible early adopters of new technologies because they are sophisticated customers, possess a base data layer to work with, and face substantial energy costs. However, at the same time, start-ups that focused on the residential market had the largest average deal size ($5.9 million versus $4.1 million for commercial and industrial). One possible reason for the attractiveness of investments in digital technologies targeted at the residential sector is the presumed future market size of the smart home, with its promise of a myriad of connected devices, battery storage, solar generation, and electric vehicles, creating opportunities for new products and services.

With respect to capital providers, the support for digital energy technology start-ups has been broad based. Focusing on seed and series A funding transactions, two hundred distinct investors participated in 126 transactions.[3] Whereas 87 percent of the investors were either VCs or CVCs for series A, the distribution of seed capital providers was more diffuse, government and incubators/accelerators representing 45 percent of participating investors. Notably, only 10 percent of these capital providers were utilities; 8.5 percent were foreign and 1.5 percent U.S.

based. This statistic suggests that utilities have been slow to capitalize on the opportunities digital innovations present and could suffer
competitively as a result. In the electricity sector, one feature of digital
energy services is the low barrier to entry: creating compelling offerings directed to customers in the form of a plug-in device, app, or online
platform often does not require the utility or a proprietary platform
to act as an intermediary. There is a proverbial race to the customer to
offer products and services, given that margins on electricity are falling
and the prospects for competition at the distribution level are increasing. U.S.-based utilities have only recently begun working with start-
ups to update their business models.

THE ROLE OF OTHER ACTORS

For digital technologies to enable a flexible, reliable, cost-effective, and
clean electricity system, support is required from a constellation of actors
broader than just VC investors. In particular, both public policymakers
and nonprofit entities can provide support that can supplement private
investment and enable start-ups to commercialize digital innovations.

The first need is expanded support for applied research into digital technologies. Government entities that support applied research
include the Advanced Research Projects Agency-Energy (ARPA-E) at
the federal level and the Electric Program Investment Charge in California. Both support the interdisciplinary research and development
needed to advance energy-specific aspects of digital technologies, such
as the ability to monitor the grid, use predictive analytics, integrate
data, and minimize the latency of digital communication. Their support
also touches cybersecurity, which is now one of the greatest strategic
concerns of utility executives.[4] Such funding programs are critical to
developing core digital technologies; these programs should therefore
be expanded.

Beyond applied research, some government programs provide crucial support for subsequent stages of commercializing new technologies. These programs include the small business innovation research
(SBIR) and small business technology transfer (STTR) funding vehicles furnished by the U.S. Department of Energy. The department
has announced funding opportunities for 2017 and 2018 that empha-
size digital innovation, making specific calls across department offices

including "advanced digital network technologies and middleware services," "sensors and control—novel concepts for blockchain-based energy systems," and "cybersecurity for solar devices." Although the competitively awarded SBIR and STTR grants should continue and perhaps expand, more could be done with existing funding by reducing the size of each grant to match the low capital requirements of digital technologies. For some digital innovations, grants in the range of $15,000 to $50,000 could inform go/no-go decisions for further development.

Finally, outside the public sector, nonprofit incubators and accelerators provide platforms for experimentation, information exchange, and mentorship, and offer an efficient forum to scrutinize ideas to reduce uncertainty in investment quality.[5] Moreover, given the velocity at which new digital technologies and their business models are coevolving, incubators and accelerators provide a window for incumbent energy companies to change their business models. The venues offer a "light-touch" opportunity for incumbents to experiment without high capital or organizational commitment. Examples include the Elemental Excelerator in partnership with Hawaiian Electric Company, and IDEO CoLab, which develops and demonstrates digital technologies applicable to both the financial and electricity industries.[6] Consortia that include utilities as well as start-ups, capital providers, and other relevant actors create more than just the opportunity to observe; they provide an impetus for strategic organizational change within utilities, whose participation is crucial if digital technologies are to be adopted at scale.

Digitalization: An Equal Opportunity Wave of Energy Innovation

David G. Victor

The energy industry is experiencing waves of innovation and more are to come. These waves have enabled profound improvements in the management of complex systems. For example, in many electric power markets, even as intermittent wind and solar electricity replaces more predictable power from conventional plants, better management of electricity systems has kept total costs from rising. This outcome reflects the confluence of many innovations. Some relate to improvements in the materials and construction of solar and wind power generators as well as energy storage equipment. Others relate to business model innovations that have opened up new supplies of capital. But the innovations at the center of the energy industry's rapid changes are digital. New sources of energy system data, such as from low-cost sensors, and the ability to harness that data thanks to increases in computational power and new data science techniques, have enabled the efficient management of complex stochastic power supplies and expanded availability of information to customers and system operators. The disruptions and transformations these innovations bring to the world economy could be as significant as those that followed the rise of electricity and oil a century ago.[1]

Barely a decade ago, the power industry was a paragon of stasis. High levels of regulation and state ownership, combined with high capital intensity and long investment cycles, meant that the future of the industry was easy to predict and not much different from the past. As a result, despite a recent flurry of investment in renewable energy, the actual penetration of clean energy technologies remains little changed from a decade ago, except in a few jurisdictions such as California, Hawaii, and parts of Germany. Recent innovations, however, could drive a clean energy revolution, especially if digital technologies make it possible to orchestrate clean and complex energy systems.

But four reasons bolster skepticism that digitalization will, on auto-pilot, advance clean energy systems at the expense of dirty ones and reduce the energy sector's overall carbon footprint—the primary pre-requisite for combating climate change:

- Rising venture capital (VC) investment in Silicon Valley start-ups peddling digital clean energy technologies may not actually herald the arrival of world-beating firms. The innovation model that has spawned big changes in information technology (IT)—and on which the future of energy is imagined—is poorly suited for the energy industry. Even though digital start-ups could improve upon the dismal performance of their predecessors—makers of new clean energy hardware—they could also fall short of rapidly transforming an industry dominated by slow-moving incumbents.

- It is unclear whether digitalization will help clean energy any more than it will dirty energy. Indeed, even though digital innovations will probably speed the deployment of renewable energy, they could plausibly benefit traditional energy sources, such as oil and gas, even more. Digital innovation, by itself, spells neither the rise of renewable sources nor the demise of fossil fuels, for the same forces that facilitate clean energy revolutions are improving every other segment of the energy industry—including fossil fuels.

- Biasing digitalization toward the promotion of cleaner energy systems will require policymakers to enact regulations that incen-tivize decarbonization, for example, by pricing carbon emissions. The problem is that no powerful lobby exists to support such policies. Instead, the renewables industry has masterfully cre-ated a political coalition that equates green energy with renew-able energy when, in fact, it is far more cost effective to generate useful energy with a small environmental footprint without rely-ing solely on renewables.

- The one certainty from the digital wave of innovation is that the future of the energy system is becoming increasingly uncertain. Digital upgrades to energy systems are important, but their over-all effect is highly unexpected. Digital innovation is allowing the energy sector to change faster than before, for example, by speed-ing decentralization. But not only will it speed decentralization of the energy system, it will also open up the energy sector to new competitors and technologies, making it harder to predict where the energy system is headed.

SILICON VALLEY IS BETTER AT SILICON THAN AT CROSSING VALLEYS

Silicon Valley, as a shorthand for the American model of IT innovation, has generated enormous and rapid changes in technology. The fundamental attributes of the model are well known: clusters of innovation sources (universities and government labs) with large network effects, easily claimed intellectual property rights, and prodigious sources of early-stage capital. These attributes have most notably helped technologies that can scale quickly and ideally become natural monopolies by providing services that nobody thought they needed.[2]

Starting about fifteen years ago, this model was replicated on energy darlings of the day—fuel cells, energy storage, and a few others. The results have been discouraging. Cash burn was high, profitability elusive. A few exceptions—such as Tesla and Stem—aside, firms have struggled to gain market share. High-return exits have been rare. From 2002 to 2012, deal flow in clean technology rose from essentially zero to about seven hundred, and total investment climbed past $7 billion. Today the numbers are about two-thirds that level.[3] Surely, some striking new developments have emerged—such as the efforts of the Breakthrough Energy Coalition, surprising advances in technologies such as small modular reactors and even nuclear fusion, and a nascent uptick in investment in digital technologies—but these are exceptions rather than the rule.

The prevailing pessimistic narrative is right. An innovation model that relies on quick scaling into monopoly will not take fundamental risks in energy systems. That is why VC investment in early-stage clean technology ventures remains flat; instead, nearly all of VC investment is concentrated on late-stage firms. Indeed, since 2008, four-fifths of all clean technology VC funding have gone to late-stage firms.[4]

The answer to this problem is funding that is patient and focused on overall social return. In some instances, governments have provided that answer. China has become the driving force in the global solar industry through massive injections of reliable capital from state financing systems and some private investment in an ecosystem in which state institutions help lower risk.[5] Although the accomplishments are huge, the vast majority of the investment in China focuses on existing technology suites; evidence is mounting that the efficiency of China's efforts to translate resources (e.g., money and people) into viable investment remains low.

Most disturbing is that traditional government sources of research and development funding have not kept pace. Substantial evidence shows that the United States has learned how to effectively spend public funds on early-stage energy-related technologies. Even so, actual funding levels remain erratic.[6] In addition, funding within the states, even as it rises, tends to focus on safe bets. For example, the vast majority of California's Energy Commission funding for advanced energy projects concentrates on deployment of known technologies, such as microgrids and storage, rather than research and innovation.

Recent data showing a creeping increase in early-stage private investment in digital technologies is indeed promising.[7] But the data does not entail the conclusion that the Silicon Valley model has finally alighted on a way to transform the energy sector after its false start a decade ago. Especially given the financial disaster resulting from the last wave of clean energy investment, it is prudent to wait and see whether this new digital wave of clean energy investment in fact bears fruit. In the meantime, the default assumption should be that the energy sector is fundamentally resistant to change, and most start-ups seeking to establish natural monopolies will run into the buzz saw of existing monopolists.

THE SYSTEMS REVOLUTION IS NOT JUST FOR GREENERY

IT is transforming complex energy systems. Renewable sources of energy are assumed to be environmentally friendly. Greener energy systems are essential to mitigating climate change. Therefore, the logical fallacy follows, the IT revolution will help energy systems fix the climate crisis. The reality is different.

IT in the broadest sense is making renewable energy systems much more capable. IT, along with improved materials, is also making it possible to build low-carbon nuclear reactors at affordable prices, at least in the markets where regulation and oversight permit so.

But IT and a host of other Silicon Valley innovations will potentially have the most profound effect on the traditional industries of oil, gas, and conventional electricity because these industries have strong existing incentives for better performance and huge potentials to realize that performance through systems integration. From 2008 to 2014, the price of oil fell by a factor of four, and given huge losses in the upstream sector, expensive oil projects were all put on hold. Since 2014, as the "lower for longer" mindset about the persistence of low oil prices became entrenched, drillers found ways to make their systems perform better. This came from improved simulation and fault analysis, better integration of subsystems, and better planning for outages, such as through predictive maintenance.[8] Control of complex combustion processes—essential for reliable power plants, including fossil-fueled ones with or without carbon capture and storage (CCS)—is plausibly poised for similar improvements.

For offshore drilling, these improvements have cut costs by perhaps one-third in just a few years. On land, a host of distinct but related changes has made horizontal drilling and fracturing of wells for shale gas and oil production much more productive and profitable. In the Permian Basin of west Texas and southeastern New Mexico—just one of the major shale oil–producing regions in the United States—oil production is steadily rising and stands at nearly three million barrels per day, even as only a tiny fraction of the resource is being tapped. Given breakeven costs of less than $30 per barrel, drillers envision a future, perhaps in a decade, of ten million barrels per day from this region alone.

Thus, the IT revolution has been an equal opportunity revolution, and though the narrative generally focuses on how it favors renewables, it is in fact shifting supply curves outward for nearly all rival sources of energy. To paraphrase the former Saudi Minister of Oil and Mineral Resources Ahmed Zaki Yamani, the Oil Age will not end for lack of oil, just as the Stone Age did not end for lack of stones.[9]

THE RENEWABLES LOBBY IS NOT A DECARBONIZATION LOBBY

For digitalization to favor energy systems with a lower carbon footprint than today's, public policies that realign the market's incentives and aim at decarbonization will be required. Yet the policies that the most powerful clean energy lobby—renewables—has pushed for would not create strong incentives for decarbonization. As a result, digitalization will continue to equally enable clean and dirty energy. Along the way, tremendous political capital will be spent lobbying for policies that are ostensibly clean but in fact do little for decarbonization.

Politically, the most extraordinary shift in the energy industry over the last two decades has been the rise of a durable lobby for renewable energy. This lobby emerged haltingly at first, but as investment in renewables grew, the lobby went mainstream and became powerful.[10] This process has been the most conspicuous for wind and solar, but in markets where biofuels play a big role—such as the United States—the biofuel lobby has emerged in similar ways. The renewables lobby has successfully advanced renewable energy, but emissions have not gone down decisively. In Germany, most famously, emissions have stayed roughly flat despite massive investment in renewable energy. Prodigious investment in lignite coal, along with reductions in nuclear power, have offset gains in renewable energy.

The renewables lobby has its challenges. The small one is that it has not been particularly adept at pushing for what is needed to make renewables actually work at scale. In the United States, the lobby is skeptical of grid expansion because that could allow competition with other energy sources (even coal), even though bigger grids are much better at integrating renewables. The lobby has, for the most part, not got serious about storage; when it has, it has focused mainly on lithium batteries, which are poorly suited to massive integration. The big issue is that all-renewables grids are probably an expensive and unreliable way to cut emissions. Almost no plausible route exists by which the renewables lobby refashions itself to allow new low-carbon entrants—such as nuclear and CCS—to become a decarbonization lobby. Politically, those sources of low-carbon energy will be viewed as rivals even though technologically they are complements; consider, for example, the lack of any broad-based interest in nuclear power within the renewables and green communities despite growing fear of climate change.

PREDICTABILITY OF THE SYSTEM HAS GONE DOWN

A system on the cusp of radical transformation is difficult to forecast.[11] Large parts of the energy value chain are becoming decentralized and open. That flat, open topology shifts authority and influence away from incumbent firms and infrastructure and toward new entrants and even consumers. Not all those entrants exist currently, so how they will change the system is difficult to anticipate.

Predictability has also declined because the waves of innovation affecting energy industries substantially change the way systems interact with one another. Each of these innovations has proved hard to predict. Together, the interactions among the systems further compound the uncertainty inherent in projecting the effects of innovation, which is partly why the track record for mainstream energy forecasting on penetration of new technologies, such as solar, has been so poor.[12] Traditionally, the inertia of the large energy industry—sheer size and capital intensity—was a source of stability and therefore predictability. However, a large system depends on continued investment. Pervasive uncertainty is making firms wary of deploying capital. In the oil and gas industry alone, by some estimates $1 trillion in previously viable projects is on hold due to uncertainty about future prices and structure of the industry. The power industry has been similarly reluctant to make capital-intensive upgrades because it is becoming unclear who will pay for the grid. New firms and policies aimed at promoting disruption have also disrupted the business models and policy credibility needed for long-term investment. Chronic underinvestment could thus lead to more intense cycles in price and behavior, creakier infrastructure, and more crisis-driven policymaking.

Most significant is that many of the sources of new ideas and technologies lie outside the energy industry. New market designs and technologies, such as distributed ledgers, arose from banking and IT. Yet these innovations could allow many more efficient peer-to-peer transactions that undermine traditional energy suppliers and marketers and erode the capacity of government to supervise and tax energy services. Renewables have emerged, in part, from advances in semiconductors; radically improved batteries stem from advances in material science and demand for power storage in computing.

These sources of innovation, for the most part, are far outside the realm of familiarity and forecasting skill for the energy industry. They tend to be less responsive to the normal market forces of supply and demand within the industry. The explosion in information and communications technologies that lies at the root of today's waves of innovation appeared on its own and did not emerge in response to changes in oil or power prices. Through 2017, Tesla, a leading disruptor in electric vehicles, had never turned a profit and had lost nearly $4 billion, even as its market capitalization exceeded $50 billion. Even profitable innovations within the industry—such as the shale boom—were not accurately spotted by many observers because they arose from niches on the periphery.

Amplifying the difficulty is that the most disruptive innovations often arise from short-lived start-ups whose survival in the search for capital and market share depends in part on hyperbole that drives valuation. The ecosystem of disruptive innovation—wherein everyone claims they will disrupt exponentially—is noisy, making it difficult to assess which ideas will survive. As in much of the Fourth Industrial Revolution, the business model for this democratic, decentralized mode of innovation tends to yield a huge churn in ideas and a few blockbusters. Success tends to be equated with prescience, when luck often plays a bigger role. And the media amplifies the problem of finding the signal in all the noise—often reporting as truth what the new class of billionaires says, without any scrutiny.

Part II: Digital Opportunities

Electric Power

A Survey of Digital Innovations for a Decentralized and Transactive Electric Power System

Lidija Sekaric

Today's electric power system is under increasing strain. Weather-related power outages between 2003 and 2012 cost more than $30 billion annually on an average.[1] The frequency of extreme weather events continues to rise, a result of climate change. In addition, as the centralized transmission and distribution grid ages, outages due to equipment and operational causes are increasing. Power losses cost grid operators, power generators, and end customers dearly; a recent power outage at the Atlanta airport cost the hub airline about $50 million. In addition, rapid customer adoption of distributed electricity generation at the grid's edges could compound the situation by adding complexity to a system that is already struggling to contain costs while providing reliable service.

But the power system's challenges are not insurmountable. Digital innovations will be central to managing the grid's existing issues as well as enabling a transition toward a new, distributed grid architecture. In the future, a digitally advanced electric power system could better withstand disasters, rely more on clean energy, and expand to meet energy needs not currently met by electricity. And, in the process, the global transformation of electricity could create jobs and boost economic output, creating $2.4 trillion of economic value over the next decade, according to the World Economic Forum.[2]

Crucially, digital technologies can enable the electricity system to evolve, step by step—it would be infeasible to abandon the existing grid infrastructure entirely and design a new grid from scratch. Over the next decade, digital technologies—such as smart sensors, predictive analytics algorithms, and digital twin models—will gradually move the grid toward a more decentralized and efficient configuration. Beyond that horizon, however, the future ubiquity of digital innovations could render the grid radically different from its present state. For example, a fully transactive grid of the future could empower prosumers (consumers of energy who also produce it) to trade electricity at the edges of the grid, recording their transactions on the blockchain. In this way, a gradual evolution could culminate in a full-blown revolution (see figure 1).

To develop and deploy digital technologies that shepherd the grid toward such a revolution, firms will need to develop new business models. To enable business model innovation, policymakers should ensure that electricity sector regulations align firms' financial incentives with improving the affordability, reliability, and sustainability of the grid.

MOVING TOWARD DISTRIBUTED ENERGY

The conventional centralized energy grid is beginning to evolve in a way comparable to the transformation in digital computing beginning in the 1970s, when desktop personal computers (PCs) were introduced. The PC supplemented traditional, centralized mainframes to extend computing power into the hands of end users. As a result, businesses and consumers that relied on computers saw significant productivity and efficiency gains at all levels. Now, when it comes to the electric power sector, centralized grid infrastructure is being supplemented by distributed microgrids, distributed renewable energy, and cogeneration systems installed by energy consumers, including corporations, municipalities, and individual homeowners. Also, just as was the case for the decentralization of computing, this shift in electricity systems has the potential to increase system efficiency and reduce costs.

Digital innovations can ensure that the centralized grid—which still accounts for the vast majority of electricity system infrastructure—works seamlessly with the new distributed equipment on the grid's edge.

FIGURE 1. EVOLUTION OF THE ELECTRIC POWER NETWORK FROM CENTRALIZED TO AUTONOMOUS

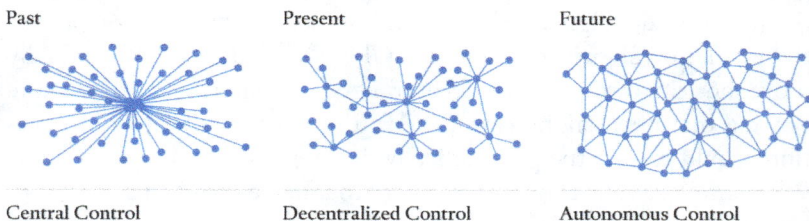

Past

Present

Future

Central Control

Decentralized Control

Autonomous Control

Source: Author.

For example, digital technologies can ensure that the deployment of distributed solar panels—a substantial portion of the increase in overall renewable electricity generation—stabilizes rather than strains the grid. Smart inverters that often come with a solar installation can help the grid cope locally with an issue such as voltage variation, if they are controlled in an intelligent way that responds to the grid's needs. Such distributed intelligence and active control helps with grid stability, counteracting the destabilizing effect of adding intermittent renewable energy to the grid.

On a larger scale, digitalization can marshal multitudes of distributed sensors, generators, and appliances to shore up the health and reliability of the centralized grid. Power utilities and grid operators can gather critical grid performance data through low-cost devices that communicate and integrate throughout the transmission and distribution networks. The result is real-time grid operational information, both technical and economic. Armed with this information, utilities can then deploy battery and other energy storage resources to meet customer needs during times of high instantaneous demand. They can also recruit thousands or millions of customers, each with internet-connected electrical devices, to shift grid energy consumption to hours of the day with lower demand, reducing the peaks in the network's demand profile. This strategy would create virtual power plants, composed of aggregated customer demand, that offer increased flexibility, such as to compensate for the massive fluctuations in power output as more renewable energy connects to the grid. In European and some other power markets, for instance, multiple distributed resources that are not owned by one entity and are located apart have been aggregated so that they can better compete in their markets. Software tools allow for much more precise analysis of power supply and demand interactions.

The rise of microgrids exemplifies a shift toward decentralization of the electric power system that does not have to be disruptive to existing centralized infrastructure if digital technologies are judiciously deployed. Modern microgrids, when controlled by digital technologies, are quite compatible and interoperable with the centralized legacy grid. Microgrids manage power generation and delivery much as the centralized grid functions. Although the concepts of controlling and balancing energy supply and demand are the same, microgrid implementations can be much more efficient and nimble. Microgrids and distributed generation use software to optimize the operation of distributed generators by identifying when such a source of supply is needed. Software can also accurately forecast the consumption of electricity by analyzing historical patterns of use. Such digital tools enable operators to continually minimize the costs or carbon footprint of the system. Microgrid customers also have the flexibility of choosing to either integrate seamlessly with the grid or to be completely self-sufficient and independent from the main grid, if desirable or necessary.

All of this is not to imply that the only benefits of digitalization are to permit decentralization and help distributed and centralized assets complement each other. In fact, digitalization also holds great promise just in the realm of the centralized electric power system, a large portion of which is likely to persist for decades. Digital technologies can enable more efficient operation of grid infrastructure—from generators to power electronics—lowering the electric power system's cost as well as its carbon footprint. For example, a digital twin is a virtual computer model of a piece of equipment that analyzes sensory data and then runs simulations to benchmark performance, allowing a grid or generation plant operator to pinpoint where efficiency gains can be made. For example, if a plant operator is considering a more efficient gas turbine, the operational engineers can use a model to optimize turbine blade design or material choice. Because a digital twin can be tested before it is even built, maximum efficiencies can be extracted from the design. Digital twins can even be used to model the operation of the power grid and account for a variety of operational stresses. Sensors can be added to existing equipment so that performance data can be gathered and patterns of data interpreted via cloud-based data analytics, leading to improved efficiency.

Finally, predictive analysis—relying on digital technologies—can help prevent failures and outages across the components of the electric power system. Digital tools can detect subtle operating variations in critical pieces of equipment that, in the past, would have gone undetected. Early detection of such faults helps greatly reduce the likelihood of catastrophic failure and costly unplanned downtime. The financial implications of predictive maintenance and downtime are also critical to financiers and equipment vendors with warranties in place.

LAYING THE GROUNDWORK FOR THE TRANSACTIVE GRID

Ultimately, the shift toward distributed energy resources enabled by digital innovations could culminate in a transactive grid, which refers to a truly decentralized network of electricity trading. More formally, transactive energy is broadly defined as "a system of economic and control mechanisms that allows the dynamic balance of supply and demand across the entire electrical infrastructure using value as a key operational parameter."[3]

The transactive energy approach promotes more interactions and transactions at all levels of generation and consumption, and offers a means for producers and consumers to more closely match and balance energy supply and demand at a certain time and place. This market-driven approach can be well suited for a grid network with a high share of distributed energy resources. It couples the physical grid with economic markets.

Grid digitalization and decentralization have accelerated the growth of prosumers: energy consumers in homes and buildings who are also energy producers, capable of supplying the grid with stored energy during periods of high demand, and who are compensated for providing access to that stored energy. In such a transactive grid, energy controls begin to play a critical role. Consider the proliferation of electric vehicles: the batteries of parked cars will be a huge resource of stored energy that can be tapped when demand is high or during an emergency.

Under such a scenario, advanced, digital energy system controls are critical. The sensors that gather large amounts of data and the cloud-based software analytics that provide the visibility and decision support needed now allow participants up and down the energy value chain to evaluate power availability, manage the distribution network, and monitor consumption.

In the future, prosumers and consumers will have multiple energy-consuming devices and assets to manage (e.g., cooling and ventilation equipment, heat pumps, photovoltaic equipment, electric vehicles, stationary batteries, and so on). To deal with these assets most efficiently, prosumers and consumers need automated monitoring and control systems that will facilitate decentralized electricity transactions.

USING NEW BUSINESS MODELS TO PROMOTE DIGITAL INNOVATION

Digitalization will enable new financial models to be built around the modernized grid that reward the parties up and down the energy distribution chain (e.g., transmission system operators, wholesale market operators, traders, distribution system operators, retailers, consumers, local market operators, distributed energy providers and aggregators, and prosumers) for making a decentralized grid more stable.

Emerging blockchain technologies offer a possible platform for enabling peer-to-peer financial transactions. Although still in exploratory stages, blockchain technology could enable the digital representation of energy and financial assets and their secure transfer from one set of parties to another. The security of this value transfer is guaranteed by the interaction protocol, eliminating the need for trusted transaction intermediaries and subsequent delays in processing. This aspect of the financing system is expected to disrupt banking, governance, and commerce practices.[4]

A smart meter participating in the blockchain network could make secure and automated trades on behalf of the consumer or prosumer whose assets it represents. The security of the transaction and the automated execution of smart contracts could reduce the cost and latency of market clearing, settlement, and billing. The direct participation of multiple parties in a smart contract can be a much simpler way of enabling consumers to pay for electricity than today's cumbersome and highly regulated process of charging and collecting rates, while blockchain systems can inherently ensure market transparency. Pilots of blockchain use in community grid as well as utility-related initiatives are already operational.

How Distribution Energy Markets Could Enable a Lean and Reliable Power System

Benjamin Hertz-Shargel

For more than a century, electric power systems have followed the same basic model: centralized and scheduled generation of electricity, followed by one-way transmission and distribution to customers. These systems have been centrally planned and operated, relegating nearly all participants to passive roles as consumers, disconnected from energy markets and management of the grid.

This model has worked quite well, ensuring affordable, reliable, and nearly universal electricity access in the developed world. But the existing model is increasingly under strain, and a radically different model might be needed to meet the new demands on power systems. The rise of intermittent renewable energy, which can help decarbonization, poses challenges to both the physical stability of the grid and financial stability of the energy markets on which it depends. Grid-scale renewables are, for the most part, not dispatchable by the system operator; this decreases the flexibility of the electricity supply and introduces uncertainty. These resources also operate at zero marginal cost, depressing the price of electricity in wholesale electricity markets, threatening conventional generation.

In addition, end users are no longer simply consumers of electricity. The declining cost and increased connectivity of distributed energy resources (DERs)—such as rooftop solar panels, stationary batteries, smart thermostats, and electric vehicles—enable retail customers to act dynamically as both consumers and producers of energy. The resulting two-way power flow, which is invisible to and uncontrollable by the system operator, makes the task of stabilizing the grid more challenging at multiple scales: At the street level, electrical pole-top voltage must remain within strict bands for customer and appliance safety; one step further from the customer, at the feeder circuit level, distribution equipment is faced with reverse power flows it was not designed to handle, as well as overload, such as from coincident electric vehicle charging; at the distribution and transmission substation, supply and demand of real and reactive power must be matched; and at the balancing area level—encompassing millions of customers and potentially thousands of square miles of territory—the frequency of the grid's alternating current must be maintained despite tremendous fluctuations in customer demand and intermittent renewable supply.

Traditionally, to balance the grid, utilities have relied on dedicated infrastructure in the form of distribution line upgrades, capacitor banks, and voltage regulators, and they have absorbed the price swings in wholesale electricity markets that they cannot control. But the cost of these investments and the low utilization factor of the distribution upgrades in particular make them inefficient and poorly suited for the impending scale challenges posed by renewables and DERs.

Digital innovations offer utilities and grid operators the opportunity to transition to a new model for the electric power system that can better and more efficiently cope with these challenges. In this model, DERs transform from liabilities into assets, the flexibility of which is harnessed by digital technologies to provide value to both the customer and the wider network. This value would be captured in a new marketplace, known as a distribution energy market, that empowers all distribution customers to contribute to—and be compensated for—grid balancing. The new model would take advantage of customers' willingness to invest in DERs and would reduce reliance on expensive, dedicated utility infrastructure funded by ratepayers. The result could be even more affordable and reliable electricity, with the added benefits of a lower carbon footprint and enhanced customer choice.

OPPORTUNITIES FOR AND BARRIERS TO INNOVATION IN A CENTRALIZED MODEL OF ELECTRICITY DISTRIBUTION

Even under the existing electric power system model, significant innovation is under way on both the customer and the utility sides. On the customer side, smart thermostats, the sole DER with significant market penetration today, are capable of learning customer behavior and the thermal properties of their homes. This capability unlocks increased energy efficiency as well as demand response, a service in which customers are compensated for allowing utilities to directly control their appliances, such as air conditioners, during peak demand periods.

Additional smart services include meter disaggregation, in which software vendors use statistical algorithms to break down a customer's building-level advanced metering infrastructure (AMI) data to the appliance level, identifying savings opportunities and encouraging conservation. Battery energy storage systems and the intelligent cloud software that manages them enable commercial and industrial customers to reduce their monthly peak electricity usage, which utilities price as a demand charge—often a substantial portion of the customer's utility bill. Even forgotten home appliances such as water heaters are being transformed into cloud-connected, flexible assets that can provide demand response and grid frequency regulation services.

At the same time, utilities are investing in software tools that make their networks more flexible, resilient, and secure. AMI analytics software is used to streamline billing, detect power theft, and inform the load models needed for distribution planning and operations. DER management systems put fleets of customer resources at the system operator's command as a virtual, distributed power plant capable of providing load and voltage relief where it is needed. More slowly, utilities are also beginning to invest in advanced distribution management systems (ADMSs), which provide a single, integrated platform for what have been isolated software systems managing meter, customer, and utility data. Despite being a considerable investment, an ADMS can improve power flow through the network and automate outage management and system restoration, enhancing resilience to disruptive weather and equipment failure. Regulatory hurdles exist in making the foregoing types of investments, but progress has been made: the National Association of Regulatory Utility Commissioners adopted a resolution in late 2016 that utilities should be permitted to capitalize—and thereby recover costs from—software-as-a-service expenditures.[1]

Despite these advances, significant challenges remain in transitioning to an efficient power distribution model in which customers share the burdens and revenue of system balancing. First, today's utility rates—which charge tiered, flat rates per kilowatt-hour (kWh) of energy consumed—do not incentivize customers to align their energy use with system needs or invest in DERs. Arizona and California have rolled out mandatory time-of-use (TOU) rates, which send coarse price signals to customers, but advanced rates are otherwise being tested only at small, pilot scales across the United States. Advanced rate rollouts are hindered not only by political and customer pressure to protect simple rate structures, but also by the low penetration of AMI, the technology that enables the timely, granular electric consumption data necessary to implement these rates. Today, fewer than half of customers have smart meters and often represent only a fraction of a utility's rate base.[2] System-wide AMI deployment, real-time access to smart meter data by customers and authorized third parties, and advanced rate offerings represent necessary, if not sufficient, conditions for customer involvement in power management.

Another challenge is that customers do not have sufficiently intelligent, integrated home energy management (HEM) tools to manage TOU or demand charge rates, let alone the real-time prices necessary for a truly dynamic grid. Although the lack of compelling advanced rates reduces the value of these tools, they have achieved limited market traction for other reasons as well. It is unclear which HEM platform, or ecosystem of platforms, will ultimately manage home energy monitoring and control. This market uncertainty has been exacerbated by numerous home area networking standards competing for market share, including Wi-Fi and its low power alternatives, 6loWPAN, ZigBee, and Z-Wave, which can be more suitable for embedded devices. Competition among the vast number of open and proprietary standards has fractured the HEM platform ecosystem, hampering device integration and the development of software capable of intelligently managing whole home energy use.

Finally, even with integrated HEM tools at their disposal, utility customers would still have limited options to leverage them in energy markets. Today's organized markets are wholesale markets that involve trading vast amounts of electricity among large actors, such as merchant generators and utilities; they were not designed for mass-market participation. Indeed, direct participation requires interconnection to the high-voltage bulk power system, excluding all but a small number of heavy industrial customers. Even without this restriction, locational marginal prices (LMPs)—the time-dependent price of electricity at a particular point in the electricity network—are defined only at transmission nodes and fail to account for distribution network flows. This means that currently no mechanism is in place to set prices at the smallest length scales of the network—the edge of the distribution system—where end customers reside and could use DERs to contribute to grid balancing.

Moreover, barriers to entry are high for participation in wholesale markets. Special hardware is required to integrate into the market's energy management system, and detailed operational data, known as telemetry, needs to be transmitted on a continual basis, with strict performance requirements and penalties. The costs associated with these requirements are simply too great for residential and small commercial customers to bear. Some attempts have been made to shoehorn retail customer DERs into markets for demand response—the aggregation of customer resources to provide peak demand relief—but those efforts have made limited progress, and demand response is currently dominated by larger industrial customers. Burdensome rules on enrollment, seasonal performance, aggregation size and participation, and customer data access make it economically and technically challenging for third-party aggregators of small-scale DERs to compete in these markets.[3]

Reforms to wholesale market rules, which are set by regional bodies regulated by the Federal Energy Regulatory Commission, could make it easier for DERs to participate in wholesale markets and would drive innovation. For example, the PJM and CAISO markets on the East and West Coast respectively could relax the rules that limit aggregations of DERs from offering demand response and flexible capacity products. Still, those markets are ahead of MISO, a market in the Midwest, which has yet to take the first step to enabling DER participation by allowing basic market access to aggregated DERs. If wholesale markets around the country take these steps, they could speed HEM innovation by incentivizing customers to invest in and then monetize DER and smart home technology.

HOW DISTRIBUTION ENERGY MARKETS COULD ENABLE A BALANCED GRID

Some utilities have articulated roadmaps of advanced rates, AMI deployment, and DER integration to forge a path toward a modern power system, but even those steps can take the sector only so far. Unless customers have direct access to markets for energy products and services, the efficiency of those markets, along with the financial and technological innovation that drives them, will remain limited. A compelling solution is an open, transparent distribution market in which residential and commercial customers can directly transact both as buyers and, for those who make investments in DERs, as sellers. The prices in such a market would directly reflect the effect of transactions on the grid, transforming physical stability from an externality into an economic objective.

Core products in a distribution market would include real power, the intuitive component of electricity that does useful work, such as power a lightbulb, and reactive power, the less-intuitive additional component required for inductive loads, such as motors and air conditioners.[4] Market services could be fashioned after those in wholesale ancillary service markets, such as frequency regulation, in which generators bid capacity to follow a real-time output signal based on current supply-demand imbalance. In the distribution market case, the signal could be based on local grid congestion, reflecting the proximity of distribution equipment to their operating limit, to which customers could respond with regulation of either demand or supply. Customers could also provide reserve power capacity to help prevent outages.

For such a market to drive the grid toward balance at both the local and system-wide level, it needs to produce granular prices—both temporally and geographically—that take into account the distributional constraints ignored by wholesale markets. Such granularity can be achieved by borrowing several constructs from wholesale market design: a day-ahead forward market, where price formation begins, and a real-time market where forward positions settle.

Several options are viable for a forward distribution market: a clearing market based on security-constrained economic dispatch (SCED), a family of algorithms used by wholesale markets, and a bilateral market, centrally managed or distributed via blockchain.[5] Regardless of type, deviations from positions held at the conclusion of the forward market would settle at geographically and temporally granular distribution locational marginal prices (DLMPs), quantities that are output by the SCED optimization along with system-wide optimal resource schedules.[6] A single authority is required to perform this nodal pricing as a collective cost minimization exercise, which means that some entity—possibly a utility—will play an important centralized role in the future power system. Decentralizing this process, such as via blockchain, would require publishing the real-time, detailed state of the electric system to all customers, a prohibitive public security risk. An advantage of a forward clearing market is that prices will more closely reflect future real-time DLMPs than bilateral markets will, given that bilateral markets are disconnected from and therefore cannot price distribution constraints. Even under a centrally managed forward market, however, opportunity remains for a secondary market hosted by a blockchain such as Ethereum, which may include core energy products as well as derivatives implemented as smart contracts.[7] Anticipating such an opportunity, numerous companies are launching their own markets today based on private blockchains.[8] Still, proving blockchain technology can incorporate external data, such as energy consumption, via so-called oracles remains a challenge.

THE FUNDAMENTAL CHALLENGES: NODAL PRICING AND MARKET COMMUNICATION

DLMPs extend wholesale LMPs in two ways that are critical for distribution markets. First, their calculation at buses within primary and secondary feeder circuits sends highly granular price signals to customers, reflecting time and location-specific system value and cost. This is the fundamental mechanism by which balancing at the grid edge can be achieved. Second, DLMPs include additive cost components for distribution-specific constraints, such as the acceptable voltage bands for distribution equipment and local circuit limitations on reverse power flow originating from customer DERs. These constraints are not priced into wholesale market LMPs for technical expedience, which is possible only by virtue of their definition on the transmission system, rather than the distribution system.[9] Like congestion prices in wholesale markets, these constraint-induced components of DLMP emerge from the SCED optimization as the marginal effects of the constraints on overall system cost.

Over a span of minutes, hours, and days, DLMP price signals would promote grid balance through price-responsive demand, storage and generation. Over the longer horizon of months and years, they would inform investor DER siting and valuation, accounting for locational needs and constraints. As an example of the former, consider peak solar insolation periods (i.e., the middle of the day), during which distributed solar power output causes local voltages to climb. As voltage constraints begin to affect the real-time pricing algorithm, the market would publish negative prices for reactive power consumption, an action that reduces voltage, prompting stationary batteries, plug-in electric vehicles, and shaded solar units to respond. (Remarkably, even if solar panels are in the shade and cannot produce real power, the inverter that translates their direct current output to alternating current can still provide reactive power to the grid, a resource with multiple benefits.) It is far more cost effective for idle customer DER capacity to respond to such incidents than less distributed, ratepayer-funded utility assets, which serve no other purpose and sit idle most of the time.

Despite their value, DLMPs present enormous challenges for the real-time market operator. They should be communicated frequently and securely to all market participants, including residential and commercial customers. The same holds for bids and offers originating on the participant side. Therefore, a dedicated wide-area communication network will be necessary, leveraging a cellular, radio frequency mesh, or powerline backbone—perhaps the utility's AMI or field area network. Cybersecurity will be paramount, and should extend to applications behind the customer's communication gateway.

On the economic side, DLMP pricing poses technical as well as computational challenges. Most wholesale markets yield LMPs that do not support the SCED solution, in that generators are incentivized to violate their assigned dispatch schedule, compromising system stability. This arises when generators submit supply curves that are non-convex—in the simplest cases, due to fixed costs or minimum output constraints. In response, markets make out-of-market payments to generators known as uplift to clear the market and ensure schedule compliance. Given the extreme degree of variability among participants and motivations expected in a distribution market, the uplift problem will likely be even more prevalent. Policymakers who regulate distribution markets will need to determine whether supporting the optimal dispatch schedule justifies making uplift payments to customers based on their declared costs, or whether such transactions are simply too problematic to include. In the latter case, the system operator would rely on a combination of regulation services, distribution equipment, and dispatchable generation to handle resulting imbalances.

What might mitigate this situation is that a methodology for carrying out SCED optimizations with provably minimum uplift payments, known as convex hull pricing, could be implemented to reduce the magnitude of the out-of-market payments.[10] This methodology has not been deployed in wholesale markets to date because of the computational challenges it poses. However, recent advances in machine learning have addressed these challenges, and applying them to the domain of distribution market pricing has begun.[11]

Finally, the limitations of real-time prices in driving efficient and stable system behavior are yet unknown. Complexity and variability in customer response is expected and could necessitate price updates on a faster timescale than incidents occur. For example, a transformer that can overload to failure over a ten-minute time frame may not be protected by a fifteen-minute market, because the constraint-induced price signal would not reach participants in time. Therefore, it is likely that distribution markets will require price updates every five minutes, consistent with wholesale markets, though doing so will only magnify the computational challenge.

AGGREGATORS: BRIDGING THE GAP
BETWEEN SYSTEM AND CUSTOMER

To manage the risks and opportunities of real-time prices, distribution customers with DERs will require smart software agents acting on their behalf. This software would be responsible for planning and optimizing energy use in a home or business, including flexible loads such as air conditioning, water heating, and electric vehicle charging, as well as dedicated storage and generation resources. Not all energy management software would be personal, however. Aggregators would play a crucial role in a distribution market, transacting on multiple end-customers' behalf and shielding them from real-time price complexities and pitfalls. They would range in scale and purpose from today's demand response providers, focused on fleet grid services, to larger retail energy providers, focused on energy resale. From the distribution system's perspective, aggregators provide unique value by addressing the asymmetry between individual and system objectives.

Even as price-takers, end customers will not have simple linear responses to prices, so no set of nodal price curves may exist that yields optimal, or even acceptable, power flows. Only aggregators can internalize system objectives and coordinate individuals to achieve them—particularly complex objectives that span time and locational scales. A typical example of such a complex, multiscale objective would be the capacity to respond to a local thermal or voltage incident while storing energy in anticipation of peak prices in several hours, all while maintaining capacity for a system-wide reserve commitment.

Enhancing their grid-balancing capabilities, aggregations of customer DERs also have useful emergent properties that their individual constituents do not. A fleet of smart thermostats could check in to their aggregation server every ten minutes, for example, due to battery limitations, preventing their individual participation in fast response reserve services. The fleet itself can provide such a service, however, albeit with a linear, ten-minute ramp, exploiting the randomness of check-ins. A fleet is analogous to a collateralized debt obligation in structured finance, whose payout stream smooths out the highly variable ones of the underlying securities.

An important challenge to fleet management will be its interaction with HEM tools, including smart software agents. This issue affects the industry today, with aggregators strategically interacting with customers through smart device interfaces, despite customer experience objectives that often diverge from those of the device manufacturer. Nevertheless, fleet and local optimization systems will need to coexist across devices, networks, and vendor platforms. Standards such as SEP 2.0 (IEEE 2030.5), selected by California for smart inverter communication, make this possible, encapsulating resource-specific objectives, capabilities and constraints behind a common energy-centric language. Effective collaboration is only in its infancy, however, and will require the opportunities of open, transparent distribution markets to fully develop.

Transportation

The Implications of Vehicle Electrification and Autonomy for Global Decarbonization

Peter Fox-Penner

The global automobile fleet is shifting from using internal combustion engines (ICE) to electric power trains. General Motors has pledged to develop twenty all-electric vehicle models by 2023, and Volkswagen plans to offer electric and hybrid versions of three hundred models by 2030.[1] Volvo leads the pack with its pledge to sell only electric or hybrid power train vehicles by 2019. This shift spells the end of the dominance of oil in world energy use and the beginning of immense challenges and opportunities for the electric power sector.

Most industry observers also believe that driverless, or autonomous, passenger vehicles will become commercially available in the 2020s.[2] These changes will likely reduce personal ownership of vehicles, end driving as a commuting chore or a career, and upend large economic sectors in advanced economies that sell, service, and insure automobiles.

These concurrent shifts toward electric autonomous vehicles will occur as the world seeks to reduce greenhouse gas (GHG) emissions to meet targets set by the Paris Agreement and potential future agreements. Currently, almost all energy used for transportation globally comes from fossil fuels—nearly 92 percent from oil and another 4 percent from natural gas.[3] The United States passed a milestone in 2016 when the transport sector's GHG emissions exceeded power sector emissions for the first time.[4] As the remaining carbon dioxide that the world can safely emit dwindles, global demand for energy to fuel travel continues to climb steadily, so the transport sector will face ever-mounting demands to lower its carbon emissions.[5]

HARNESSING VEHICLE ELECTRIFICATION TO REDUCE CARBON EMISSIONS

Vehicle electrification can best reduce carbon emissions if the power sector simultaneously transitions from fossil-fueled power plants to

cleaner electricity generation sources. Even without a clean power transition, though, electric vehicles (EVs) would be an improvement from a
decarbonization perspective. EVs are about three times as energy efficient as ICE cars; a net reduction in GHG emissions per mile requires
power that generates less than 2.35 pounds of carbon dioxide per kilowatt-hour (CO_2 per kWh), a standard nearly every electric system in the
world already meets.[6] Further carbon emission reductions will require
a transition to a fully carbon-free grid alongside a shift to EVs for both
new and replacement cars.

Current trends are highly encouraging. Several experts now predict
that unsubsidized wind and solar power will become the least expensive
raw electricity sources in every region of the world.[7] Electricity storage, which is essential to managing variability in renewable energy on
power grids, is also getting cheaper rapidly. And as battery costs decline
and EV sales increase, EVs are expected to cost the same as ICE vehicles
within the coming decade, signaling a tipping point in customer acceptance of EVs.

For these trends to continue, policymakers and industry leaders need
to continue research and development of vehicles, systems, and batteries and support investment in EV charging infrastructure. The incon-
venience of recharging an EV or the inability to do so, which causes
so-called range anxiety, discourages many people from buying an EV
today. A nationwide EV charging infrastructure in the United States
requires a minimum of 8,500 public fast-charging stations, many of
which cannot provide attractive rates of return to commercial investors until the EV fleet gets larger and the stations can generate more
revenue.[8] Finally, as EV sales rise and EVs approach cost parity with
ICE vehicles, EV purchase subsidies for mainstream light-duty vehicle
buyers should be phased out.

Beyond their benefits from the perspective of decarbonization, electric vehicles present the electric power industry with a set of risks and
opportunities. Charging an EV can require power flows one-third to
one-half as large as that of a typical U.S. house. Many of the four million miles of electric distribution circuits in the United States (or those
elsewhere in the world) are not sized to accommodate a rapid influx of
electric loads this large; they will have to be reengineered at substantial costs and then managed in order to avoid other system problems. In
the future, utilities would also need to upgrade the grid to charge whole
fleets of corporate-owned EVs concentrated in a small geographic area,

raising economic and technical challenges. It is unclear how utilities will recoup the costs of this infrastructure and the technical capability to integrate large numbers of EVs.[9]

Of course, vehicle electrification also presents a major opportunity for utilities, which have seen largely flat and sometimes declining electricity sales for nearly a decade. By 2030, transportation will be the largest new source of electricity sales growth.[10] By 2050, U.S. power use will increase by roughly 25 percent or more simply to power cars and small trucks; large trucks, light rail, ships, and even air travel will increase sales even more.[11]

Because utilities are regulated by policymakers, states could naturally view the advent of EVs and their attendant complexities with apprehension, especially because regulators need to simultaneously contend with overseeing the digitalization of the power sector, the integration of variable renewable sources and decentralized generation, and new industry business and regulatory models. But EVs need not simply be considered as electric loads that will strain the grid. Rather, they could represent sources of electricity storage and load management as well as regulation of grid frequency and voltage to ensure reliability. In other words, EVs can provide value to a well-orchestrated smart grid. This possibility increases the rationale for investment in a charging infrastructure that is intelligent and well integrated with the rest of the power system. In tandem, it is important for regulators to change electricity rate structures, such as by introducing dynamic pricing, to give customers incentives to use their EVs in ways that help the rest of the power system.

ACKNOWLEDGING THE UNCERTAIN EFFECTS OF VEHICLE AUTONOMY ON CARBON EMISSIONS

Autonomous vehicles (AVs), if they are not fully electric, could increase emissions. For example, AVs are widely expected to increase travel by commuters, business travelers, and underserved populations such as the elderly and the disabled.[12] AVs may also travel empty to pick up the kids, cruise the block if parking is not available, or even become constantly roving platforms for commerce—"Starbucks on Wheels," in the words of one futurist.[13]

By some estimates, automated taxis will become so cheap that travel will shift from walking, biking, and mass transit use to on-demand AVs.[14] Current nonautomated forms of on-demand ride services are already estimated to have caused a decline in New York City subway ridership and increased traffic congestion in Manhattan.[15] These trends might only be the tip of the iceberg. Some experts predict that AVs will double U.S. vehicle miles traveled, doubling transport energy and potentially carbon emissions. Even with much lower estimates of increased travel, many experts think that AVs will increase congestion and urban sprawl, at least in the coming years.[16]

From an emissions standpoint, increases in vehicle miles traveled (VMT), congestion, and sprawl could be troublesome. If AVs use internal combustion engines, the increases in VMT will boost carbon emitted per vehicle well beyond the levels predicted by existing forecasts. Meanwhile, increased travel by electric AVs could take place quite rapidly, so any reduction of emissions from switching from ICE vehicles to EVs and increasing the penetration of clean electricity sources might be overtaken by the emissions increase from more driving.

Not everybody agrees that AVs will increase emissions. Many experts and even some auto executives believe that cheap autonomous taxis will prompt households to stop owning vehicles.[17] They also posit that AVs will accelerate interest in vehicle sharing and ride pooling, which could reduce rather than increase VMT. Some data supports this view, but no consensus prevails currently. It is probably safer to conclude that Americans will actually travel more as the time and cost of travel decline, and that the average number of people in a vehicle per mile traveled will not change much. As a result, even though sharing and pooling become extremely important to the auto industry, urban lifestyles, and the urban landscape, they would be unlikely to change the overall carbon footprint of the transportation sector.

Other aspects of AVs will reduce energy use—though probably not until the 2040s at the earliest —and thus could reduce the carbon footprint of the transportation sector. AVs have the potential to operate much more efficiently than their nonautonomous counterparts, with two important provisos. First, many of these advantages require that the AVs be continuously connected to smart traffic management systems (hence known as connected AVs or CAVs).[18] Second, the concentration of AVs in each section of traffic needs to be sufficiently large—that is, the number of driver-operated cars needs to be so small that human error does not foul up the autonomous traffic management or cause collisions.

Once these conditions are met, CAVs can be managed so as to reduce energy use dramatically. They will follow one another closely on roads, increasing throughput and allowing for platooning, which reduces aerodynamic drag. Braking and accelerating, which together account for the majority of auto energy use, will be reduced significantly. Accidents, which account for one-fifth of all traffic congestion, should be all but eliminated, allowing AVs to be built of lightweight materials that require much less energy to move. Conventional stoplights could become a twentieth-century relic, replaced by smart intersections where AVs glide by each other without stopping. Hunting for parking will also end.

In summary, the effect of AVs on carbon emissions is highly uncertain: it is not even clear whether AVs will increase or decrease carbon emissions, let alone by how much. What is clear, however, is that for AVs to decrease emissions, they should be operated in a coordinated fashion, be powered largely by electricity rather than fossil fuels, and replace nonautomated cars to reduce system inefficiencies. If these conditions are met, AVs have great potential to reduce carbon emissions and also to improve quality of life in cities. Moreover, all this hinges on the development of fully driverless technology, consumer acceptance of AVs, and, above all, the state of AV infrastructure.

BUILDING—AND PAYING FOR— AUTONOMOUS VEHICLE INFRASTRUCTURE

Current developers of AVs are highly focused on creating vehicles capable of operating safely without any changes to the current U.S. roadway system. This seems to have lulled much of the transport community into thinking that little or no infrastructure investment will be required to integrate AVs into the system.

This conclusion is premature: AVs will need substantial digital and physical infrastructure to thrive. One oft-overlooked need is that of investing in the computational processing infrastructure—both on board vehicles as well as for the central systems that will coordinate AV fleets—to efficiently manage AV technology. According to Intel, one CAV may need to process terabytes of data every hour, a feat that will also require processing in parallel with thousands of counter-parts.[19] Managing AVs will also require new electronic signage and control technologies. The first commercially available fully driverless vehicles may be restricted to AV-only areas that need to be designed, built, and integrated into the rest of the system. Even if early AVs do not require their own roads, policymakers could feel pressed to create AV-only infrastructure to allow the operating benefits of these vehicles to be realized in a limited environment. For example, AV-only parking facilities will allow for more remote and compact vehicle storage, but they will have to operate alongside conventional parking as long as both types of vehicles are in use.

To counteract a large shift away from mass transit to cheap auton-omous taxis, transit policymakers are starting to experiment with seamless mobility systems (SMSs). These systems use advanced scheduling and management software to enable riders to move quickly from a taxi or minibus near their point of origin to high-capacity tran-sit and then back to another taxi or bus for the so-called last mile to their destination. SMSs can lower energy use and improve the urban environment, but their implementation could be stymied by a short-age of funds. Many cities around the world are struggling with press-ing economic and social challenges and already underfunded transit systems. It is difficult to imagine the New York City transit system, which still uses analog switches from the 1920s, shifting suddenly to a highly intelligent SMS.

Many urban design experts and environmentalists believe that, ulti-mately, AVs will use roadways and parking so much more efficiently that cities will be able to eliminate many roads and use that land for build-ings and green space instead. This may be the case, but at least in the United States no widespread mechanisms are in place for integrating transport changes of this magnitude with land-use planning and public or private financing.

Remarkably, virtually no estimates of the cost of roadway upgrades needed to accommodate AVs have been published. Some pilot pro-grams are under way, however. South Korea, which spent slightly more than $12 billion on annual transport investment and maintenance in 2013, has recently announced its intention to retrofit its entire highway system at an estimated cost of about $62 billion. In the United States, Ohio will reportedly spend about $2 billion to retrofit "smart mobility corridors," including a thirty-five-mile section of a four-lane highway.[20]

These examples immediately raise the question of how cities, regions, and countries will pay for such infrastructure. U.S. road infrastructure currently holds a grade of D and has an estimated backlog of $836 billion in highway and bridge capital needs.[21] The Highway Trust Fund is estimated to reach a deficit of $80 billion by 2025, and half of all roadway outlays are now paid from general tax revenues, not user charges.[22] It is likely that the AV revolution will trigger a reexamination of roadway financing, probably culminating in a much wider use of automated roadway pricing.[23]

OVERDRIVING OUR HEADLIGHTS

The electrification of passenger transport is an essential element of global climate policy. Although it raises many operational challenges for the grid, it is highly synergistic with the industry's overall digital transformation. For better or for worse, the infrastructure needed for EVs is an increment to the existing grid that requires little change in the current urban built environment. Conventional EVs will lower GHG emissions dramatically as the grid itself decarbonizes but otherwise will do little to change urban lifestyles or energy and land-use patterns.

As part of the EV transition, policymakers will need to cope with dramatic changes in national trade and security, accelerate progress toward carbonless power systems, and ensure that digitalization and decarbonization leave the power system affordable, reliable, and universally accessible. These are immense challenges, but they are also opportunities to save the world's climate, extend the enormous benefits of electricity access to everyone, and invest trillions of dollars in a clean energy future.

The emergence of autonomous vehicles adds another large layer of disruption and policy complexity to this picture. Much more work is needed to understand future scenarios and guide the industry toward the ones that reduce, rather than increase, emissions. Much better data is needed on the realistic changes needed to make to U.S. road and communications infrastructure to accommodate AVs at each penetration level, and how these changes can be staged so they need not be completely redone as the AV fleet grows. Better data is also needed on how these vehicles will coexist with conventionally driven cars and trucks and how efficiency and safety improvements can be accelerated in the presence of mixed fleets. Finally, almost no data is tabulated on how much the infrastructure changes for AVs will cost, much less on how they will be financed.[24]

The vast infrastructure developed for driver-operated vehicles refashioned the landscape and lifestyles of the developed world. Total mobility improved immeasurably, but it came with enormous costs from environmental and security ills to urban decay, unequal access, and reduced family health. AVs are likely to retransform the urban built environment, but it will be up to policymakers to guide the changes so that they improve rather than exacerbate the effects of the last transport buildout. It will be a busy era for transport and energy policymakers, alive with opportunity and fraught with risk.

Autonomous Vehicles and Cities: Expectations, Uncertainties, and Policy Choices

Rohit T. Aggarwala

Autonomous vehicles (AVs) have gone from the realm of science fiction to being one of the most-discussed topics in technology, transportation, and urban planning. In August 2017, the research firm Gartner issued its annual "hype cycle for emerging technologies" and placed AVs at the top of the "peak of inflated expectations," just at the point where it believes a technology descends into the "trough of disillusionment."[1] At the same time, not all of the expectations for AVs are hopeful. Zipcar founder Robin Chase wrote an article in 2016 entitled "Self-Driving Cars Will Improve Our Cities. If They Don't Ruin Them."[2] And the timeline for when self-driving technology will become common on streets is highly uncertain. One indicator is that Waymo (the Alphabet company focused on self-driving vehicles) already has self-driving taxis in test service with "regular" passengers in Phoenix, Arizona, and has announced plans to purchase a fleet that would have the capacity to provide a million rides every day in 2020; by contrast, Uber's CEO recently declared that "full autonomy" on the roads is ten to fifteen years away, and the pedestrian fatality in Arizona on March 18, 2018, cast further uncertainty on the timeline of AV adoption.[3]

It is clear—and widely accepted—that self-driving technology is feasible and forthcoming, and that it will significantly change how cities and economies function and how human activity affects the planet. What is also clear, but less widely accepted, is that significant policy decisions need to be made soon to shape self-driving technology into a constructive force for cities and the planet. Fully self-driving technology will, sooner or later, be available at an inexpensive price point. What is uncertain is how policies, regulations, business models, and user preferences will shape the integration of that technology into how cities function and whether their carbon footprints expand or shrink. Indeed, the wide range of scenarios for how AVs could increase or decrease vehicle miles traveled, inhibit or encourage electric vehicle adoption, and compete with or complement public transportation creates high uncertainty over whether this digital innovation will advance decarbonization.

EXPECTATIONS

It is highly likely that AVs will be deployed as taxis, beyond testing and trials, in a few cities in the next five years and commercial sales of fully automated vehicles will follow soon after that. General Motors has applied to use cars without steering wheels in taxi services similar to Waymo's commercial services in 2019.[4] Commercial sales of fully autonomous vehicles, however, are probably still five to ten years away; the bill of materials is still high, and to date a majority of the companies developing AV technology have expressed a preference for maintaining vertical integration of developing the vehicles and operating them, both to ensure good maintenance and operation and also, no doubt, to protect trade secrets.[5] It is, however, also rational to expect that low-cost competitors will follow the first entrants within a few years and start selling AVs to end customers.

AVs will improve overall safety on the roads, at least after a certain penetration level. The safety benefits of AVs are obvious: assuming reliable operation, computers will not fall asleep at the wheel, get enraged at other drivers, or read emails and text messages while driving. In a future in which most vehicles are autonomous, the number of crashes would likely fall by 60 to 95 percent.[6] However, in the interim period, as AVs remain a minority of the vehicles on the road, whether AVs in mixed traffic will increase safety remains unclear: differences in driving styles between AVs and human drivers could lead to more accidents.

AVs could make eliminating curbside parking possible but will still need off-site parking and maintenance areas, and curbs would need to be more precisely managed as high-volume pick-up and drop-off areas. The ability of the car to continue without a driver and pick up a passenger on demand would eliminate the need to park close to the destination; in short, any place would have the equivalent of valet parking. On-street parking could be eliminated, resulting in an improved experience for pedestrians and cyclists, as well as increasing throughput. However, AVs will still require three kinds of storage:

- *Staging areas.* It is unlikely that shared-ride vehicles in even the most densely populated areas will be constantly in use (New York City taxis cruise empty 31 percent of the time during the day), and peak and off-peak times for rides will persist. AVs in cities will require taxi stands or staging areas, such as those many airports make available. Although these stands will not need to be at the front door, they will need to be only a short ride (approximately three minutes) away for the passenger to perceive no delay.[7]

- *Storage and maintenance facilities.* Shared AVs will likely need the same level of intensive cleaning and maintenance that taxis and other shared vehicles do today; as a result, taxi bases on the edge of business districts will likely continue to be needed.

- *Off-site parking areas for private AVs.* Except in jurisdictions that mandate all-shared vehicles, some cars in most cities will inevitably be privately owned and will require parking space close to the owners.

Not all AVs will be electric vehicles (EVs) from the start; the two technologies will evolve alongside each other and eventually merge. In the visionary future, AVs would all be clean electric vehicles, and if AV adoption were to hasten EV adoption, the environment would clearly benefit. However, autonomous steering does not appear to have a necessary relationship with propulsion fuel. Currently, only General Motors and Tesla are working on electric-powered AVs; Waymo and others are using converted gasoline-powered vehicles. In fact, the power draw of the guidance and computing systems of fully self-driving systems could overtax existing EV batteries.[8] As a result, it is reasonable to expect that a shift of AVs from petroleum to electric platforms will take place only gradually as battery performance improves and costs decline, and as the software and detection systems of AVs become more efficient.

In some projections of AV adoption in cities, AVs will replace public transit more than they will replace private vehicles, resulting both in the decline of transit ridership and an increase in traffic congestion. However, a rational expectation is that AVs, operating as shared taxis, will compete with transit mainly on those routes where transit is unattractive, such as low-volume bus routes serving low-density populations. AV proliferation could also lead to some drop-off in high-density transit ridership during off-peak hours, mainly at night. Still, AVs will alter but not replace public transit. The reasons are straightforward. In dense cities, even shared vehicles (four to six passengers) cannot replace the throughput of subway, streetcar, bus, and rapid transit services, because it is unlikely that streets can be expanded in those areas, and the demand for transportation will outstrip the potential for AVs to increase throughput through better driving. Further, the logic of shared-ride services in high-density areas will quickly lead to transit-like services. Consider that if everyone were to subscribe to a carpooling service, the coincidental demand for trips—say, from the Upper East Side to Grand Central Terminal—would prompt the carpool company to use a vehicle that could carry twenty to thirty people every five minutes. In short, it would look like a highly frequent bus. At the same time, transit operators would need to scale back services that are, in fact, better served by AV taxis or would face ridership drops with no corresponding decline in costs.

The likelier effect of AVs on transit, therefore, will be in reinforcing the ridership of existing, high-speed, high-density routes and undermining the ridership (and economics) of low-frequency or low-speed routes.[9] This is not necessarily bad for transit; Jarrett Walker's redesign of the Houston bus system is predicated on the idea that transit is most successful when service is provided frequently on routes with high density.[10] Further, the potential for AV service to provide attractive feeder services could make new rail lines more attractive to build. Moreover, if shared AVs were to provide high-quality, low-cost service during the night, when most urban roads are free-flowing, subways could be closed for nighttime maintenance, thus helping improve service and reduce costs overall.[11]

AVs will increase throughput but not by revolutionary levels. The potential for AVs to travel more closely together safely has been identified as one of their strong features: so-called platooning could increase road throughput. However, how far this can go is limited; in a mixed urban environment of pedestrian and bike traffic, high-speed, tightly packed AVs would create an even more hostile environment for pedestrians than current urban streets do.

AVs will not eliminate traffic signals or enable free-flowing intersections. The potential for vehicle-to-vehicle communication has led many to imagine a world in which traffic lights are abolished and vehicles coordinate so that they can move without stopping at intersections. However, the presence of pedestrians would ensure that, even in an all-AV world, some system of visual signaling would remain a part of the streetscape and cars would need to stop at various points during their journeys, at least in high-density, walkable areas.

UNCERTAINTIES

The timing for when government rules will permit AVs on the road is highly uncertain. Given the attention paid to AV technology, several countries—including the United States—will likely adopt basic licensing rules for AVs, probably at national levels; however, the speed with which these are adopted could leave open significant gaps in testing, reliability, and details. It is unclear whether governments have the capacity to evaluate the performance of a vehicle that relies on computer code.[12] But, in the wake of the March 18, 2018, death of a pedestrian, the need for governments to be able to evaluate the performance of self-driving technology has become increasingly clear.[13]

Although liability has often been raised as an issue with AVs, the paths to litigation are clear. What is far less clear is whether early crashes, failures in code, or intentional violations found in code or sensing systems could result in consumers rejecting the technology. A backlash against AVs taking over the roads is possible if they drive in a way that human drivers and pedestrians find unacceptable.

AVs will likely reduce the availability of jobs, but the extent to which jobs will be eliminated is unclear. Several commentators have noted that driving is one of the largest employment opportunities in the United States today, and one that is open to relatively unskilled labor. As driving jobs, as they exist, are eliminated, it is unclear how many jobs would either remain or be created to cover tasks that are currently performed by a driver. For example, it is possible that a platoon of driverless trucks would have a "rider" traveling with the shipment; this might mean a 16 to 33 percent loss in jobs but not a complete loss. Similarly, even as delivery vehicles become autonomous, the task of handing off a package to a person could still require a human attendant; in this case, the job loss would not be significant. At the same time, it is unclear what new jobs AVs would create. Certainly if the industry moves to more shared-ride, high-usage fleets, the maintenance and cleaning needed for those vehicles will increase, thus creating jobs.[14]

New business models would likely change road usage patterns, but the nature of the change is unknown. The fundamental economic fact of the AV is that it does not require the labor of a driver; this change in economics will therefore likely give rise to business models that do not currently exist. Robin Chase has conjectured that Amazon or its competitors could choose to pack the roads with mobile warehouses that keep driving around so as to offer guaranteed five-minute deliveries for items they expect to be purchased in a neighborhood.[15] If capital is the only constraint on placing shared-ride vehicles on the road and wait time the driver of consumer choice, then rich firms will likely attempt to saturate critical areas with vehicles, achieving location preemption, much as retail, drug, and coffee chains are believed to be doing currently.[16]

The continued attraction of the private automobile remains the greatest uncertainty for AVs in cities. Although some experts have promoted the idea that people—especially millennials—would live an "asset-light" existence and thus gravitate toward shared-ride services rather than own a car, private AV ownership will likely remain an option. Models suggest that in lower-density cities, any switch to ride-sharing will increase overall travel time for people who now drive. If, as seems likely, AV technology becomes available for purchase, it is entirely possible that a large portion of the population will simply own an AV instead of a driver-operated car.[17]

POLICY CHOICES

Despite the uncertainties, it is clear that policy choices—at the city, state, federal, and international levels—will strongly influence the way AVs are adopted and used and the ways in which AVs will change cities and the planet. These include the following:

- *Establishing standards for safe driving.* Assuming that governments figure out how to evaluate and certify vehicles that meet their standards for safe operation, the issue of "how safe is safe" will likely be politicized. Questions, for example, could be raised about whether a passenger who wants to get to a destination faster should be able to choose an aggressive driving mode, or whether every vehicle must stop, every time, for a person crossing the street despite a do-not-walk sign. Although the trolley problem—the AV deciding whether to protect its occupants or outsiders in the event of a crash—is unlikely to be a frequent occurrence, a more frequent occurrence would be the vehicle deciding how much of a threat it can intentionally create to secure its right-of-way.

- *Deciding whether, how, and when to redesign streets entirely.* Safe-driving AVs offer the potential for city governments to redesign streets to make them more pleasant and usable for pedestrians and cyclists, as well as for non-transportation purposes. However, in most places, AVs will likely be adopted gradually and operate on existing streets. Policymakers will need to decide on when, and how aggressively, to remake streets to capture the safety and other benefits of the AV for non-passengers; many current advocates for pedestrians and cyclists could quickly advocate for excluding driver-operated vehicles and accelerating the switch to a full AV city (or world).

- *Deciding whether to implement road user charging.* Roads are over-used in congested metropolitan areas because they are free to drive on; only a few cities around the world have adopted road user charging for urban streets, owing to a widespread belief that a right to the road exists. However, such a right might not intuitively apply to a machine traveling unoccupied, which is more like a vending machine or billboard placed on a city sidewalk. As a result, increased AV use offers city and state governments the opportunity to implement usage charges. If done well, pricing regimes could shape AV usage patterns and foster the kind of shared-ride, non-private-ownership paradigm that could reduce traffic, fuel consumption, air pollution, and greenhouse gas emissions in major cities. Whether this happens will depend almost entirely on local politics.

Harnessing Big Data and Data Science Across Energy Sectors

How Data Science Can Enable the Evolution of Energy Systems

Kyle Bradbury

Over the last decade, the availability of data about energy systems has surged, and in parallel advances in machine learning techniques to analyze that data have been rapid. The confluence of these two trends could reshape the energy industry. In particular, data science could enable the decentralization of the centralized energy systems that have existed since the beginning of the twentieth century. Such new system configurations, unlocked by data science techniques, could fundamentally change how societies generate, manage, and consume energy, by increasing system efficiency and optimizing planning to reduce system costs and environmental damage.

THE RISE OF ENERGY DATA: EXAMPLES FROM THE ELECTRIC POWER SECTOR

The twentieth-century conception of electric power systems is of a centralized utility operating a system designed for the flow of power from massive electric generators to end users. In this setting, utility success was sometimes defined by how little the customers thought about their electricity service. This model is now being transformed by emerging technological capabilities, particularly those innovations that are using emerging sources of energy system data. Over the past decade, new data from electricity systems has included improved power grid monitoring data, richer customer billing data, and a host of Internet of Things (IoT) data sources.

The way the operation and health of bulk power grids are measured has dramatically changed. Traditionally, sensors that were part of supervisory control and data acquisition (SCADA) systems monitored transmission line voltage and current to provide system operators with the information needed to maintain balance between electricity supply and demand. Twentieth-century SCADA systems typically collected one measurement every two to four seconds. Modern phasor measurement units (PMUs) enable thirty measurements per second, and these measurements are time-stamped so they can be compared across the grid. This provides significantly more data (up to 120 times more) to monitor the health of the grid and automate systems to detect and remediate system faults to maximize reliability. In addition to physical system sensors, in areas of the United States where electricity is traded on wholesale power markets, sources of pricing data from markets are expanding and include real-time and day-ahead energy use; frequency regulation, voltage regulation, and reserve capacity; demand response; forward capacity; and transmission congestion.

End-user electricity consumption data has also greatly expanded. For most of the last century, end users of electricity received a utility bill with one monthly data point summarizing their consumption. In 2007, seven million (6 percent) U.S. households had smart meters. Over the next decade, around sixty-five million smart meters were deployed, representing more than half of all 126 million households.[1] Smart meters provide 2,920 times more data points than monthly manual meter readings. These higher-frequency data streams enable advanced customer rate structures, including time-of-use and real-time pricing, as well as numerous previously infeasible end-user applications, such as modulating energy consumption from individual appliances to save energy at critical times for the grid. Indeed, customer devices that consume electricity, such as thermostats, are increasingly connected to the internet—adding to the growing population of IoT devices—and can provide information on energy consumption behavior and consumer preferences.

Outside the electric power system, numerous additional sources of energy system data have emerged. Vehicles are also beginning to produce data from sensors that measure driving performance and fuel efficiency, and electric vehicles (EVs) additionally provide data on their battery status. Even remote sensing data, particularly satellite imagery, has begun to be viewed as a source of information on energy systems that are visible from space.

PARALLEL ADVANCES IN DATA SCIENCE TECHNIQUES

Generally, these emerging sources of energy system data cannot be analyzed manually because of the volume or streaming rate of the data. Automated analysis techniques are required to extract the valuable decision-informing insights the data contains. Fortunately, the rise in sources of energy data has coincided with significant advances in data science and computational processing power.

Data science is an interdisciplinary umbrella term that describes applying the combined toolkit of statistics, mathematics, and computer science to extract value from data through description, inference, prediction, and strategy development (the description, prediction, and strategy components are often grouped together as machine learning).

Descriptive tools help summarize and visualize data effectively. For example, an unsupervised learning algorithm might group data into meaningful categories without any prior human input. Inferential tools provide a structured way to ask questions of and draw conclusions based on data. Predictive tools enable learning from past examples to make predictions. For example, a supervised learning algorithm might learn from a dataset of human-labeled images to predict what object a new image portrays. Finally, strategic tools learn which actions to take in new situations to optimize the outcome through interacting with the environment; for example, reinforcement learning algorithms play board games by learning which moves lead to good or bad outcomes. To use these four types of data science tools, one needs to be able to manage large datasets, also known as big data. Large database and file system management platforms—tools from a distinct field of data engineering from data science—enable machine learning techniques to work on the data.

Both the performance of data science techniques and the computational power to enable them have grown rapidly in the last half decade. Since 2012, deep neural networks coupled with powerful graphics processing units have enabled seismic changes in machine learning tools. Most notably, this is seen through image classification and segmentation vis-à-vis the ImageNet competition in which neural networks proved their predictive power on a massive database of semantically labeled imagery.[2] Additionally, reinforcement learning techniques have been successful to the extent of defeating expert human challengers at complex games of strategy such as chess and Go.

Another major change is that many advanced data science and machine learning techniques are no longer the preserve of the academy or the top commercial research establishments. Far from being obtuse code written by specialists for specialists, many proven machine learning techniques are now available through open-source repositories, a prime example being TensorFlow. This democratization of data science enables organizations without massive teams of machine learning experts to benefit from these tools.

APPLICATIONS OF DATA SCIENCE IN THE ELECTRICITY SECTOR

Given the growth in data availability in the electricity sector, machine learning techniques have the potential to revolutionize electric power systems by enhancing system operation, forecasting, and planning.

ELECTRICITY GENERATION AND DELIVERY

As clean, renewable energy sources—such as wind and solar—are increasingly integrated into power generation, uncertainty in electricity supply is increasing for the first time since the inception of the power grid. This uncertainty necessitates improvements in forecasting the production of wind and solar power as well as new market designs that can efficiently integrate these clean, low-marginal-cost resources into power generation and thus help avoid their curtailment. Machine learning approaches can use historical wind and solar data to improve forecasts. Additionally, machine learning techniques can simultaneously improve the prediction of customer demand, and therefore make it possible to more efficiently operate electricity markets that determine the schedule of power generation units, known as unit commitment.

Data can benefit not only new but also existing generation technologies. For example, reinforcement learning techniques could learn to operate components of a fossil-fueled power plant differently than human operators, creating strategies to adjust system controls that automatically optimize the reliability, greenhouse gas emissions, and fuel consumption of power plants.

An overwhelming portion of power in most countries is generated by large power plants, but distributed generation is also growing significantly. Distributed generation data is not dispatched by grid system operators and instead are allowed to inject power into the grid whenever it is produced. Small-scale solar power for the first time actualizes a two-way flow of power on distribution lines, making the modeling and management of power flows far more complicated. Small-scale distributed solar photovoltaic capacity in the United States nearly doubled in two years, from 7.3 gigawatts (GW) in 2014 to 13.2 GW in 2016. Utility scale solar grew even faster during the same period, from 8.7 GW to 19.7 GW.[3]

As distributed generation grows, so will the uncertainty in the estimates of how much electricity the grid needs to deliver to end users. A decade or more from now, distributed energy resources might reduce overall demands on the grid, for example by enabling locally managed microgrids that require minimal support from the central grid or the aggregation of large amounts of customer demand into virtual power plants that can help the grid balance supply and demand just as conventional power plants do.

But even though such innovations will take place on a small scale in the coming years, distributed energy resources will probably first impose greater demands on system operators before they ultimately provide relief in excess of strain. Ensuring reliability by predicting (or rapidly responding to) rare, high-impact events will require adaptive statistical models to forecast customer demand curves that include information on PMU readings of current and voltage and on generator outages and weather patterns, including wind forecasts and meteorological data related to cloud cover for solar estimation. System operators could even incorporate highly distributed data sources into their decision-making toolbox, including information from individual building smart meters and geotagged social media data, both of which may contain non-definitive but suggestive information about the health of the grid.

With growth in power system uncertainty, the potential value of energy storage and optimized operation of distributed energy resources increases. Energy storage could eliminate the need to simultaneously generate and consume electricity, which could reduce the use of costly peaking plants. However, for energy storage systems to be most effective, their placement within the power grid needs to be carefully selected based on grid topology data and real-time operational data. The operation of energy storage systems could be jointly optimized to rebalance supply and demand while avoiding any increase in transmission line congestion.

ELECTRICITY CONSUMPTION

Smart meters are a rich source of data on end-use consumption and numerous applications. Data on building-level energy consumption informs changes to the utility load curves, vital for capacity planning. Additionally, building-level smart meter data enables time-of-use-pricing tariffs, demand charge management and reduction, and improved nontechnical loss detection (i.e., theft). However, even more information could be revealed when predictive algorithms are applied to building-level smart meter data. The building-level energy data may be decomposed into appliance-level information through a process referred to as nonintrusive load monitoring, or energy disaggregation.[4] This technique uses supervised learning to identify which appliances are in use in a building and how much energy they are consuming. The obvious application of such a technique is to automate building energy audits. However, this technology has numerous other applications. Device-level energy consumption data could be used to predict equipment failure, measure and verify demand response, and detect opportunities for improvements to building energy efficiency. Utilities could use this information for market segmentation to better understand and meet the needs of customers and for improved estimates of appliance-specific load curves for long-term growth predictions.

Building automation could be greatly advanced through these energy data analysis techniques. And IoT devices, such as smart thermostats and appliances, could provide the data for improved effective home energy management and control systems. This information could be used either for energy use feedback to encourage energy-efficient behavioral change or for automation. Automated control systems coupled with time-of-use pricing could further add value for both the customer (by enabling an energy arbitrage opportunity) and the system operator (by providing peak shifting).

Energy storage systems are a growing alternative option in automated building energy system management. Energy storage deployment beyond traditional pumped hydroelectric storage has grown significantly in the past few years as the cost of lithium-ion battery storage has dropped precipitously. The Tesla Powerwall lithium-ion energy storage system was one of the first commercially viable batteries designed specifically for home energy storage.

In most buildings, however, other sources of energy storage exist that have yet to be fully harnessed. Essentially, all thermal energy systems in a building store energy that is slowly dissipated over time. For example, heating, ventilation, and air conditioning (HVAC) systems heat the air in a building in winter for thermal comfort; if the air were heated to a higher temperature earlier, a properly controlled HVAC system would be able to shift energy consumption from peak hours to off-peak hours. Similarly, other thermal comfort systems, such as water heaters, could also store electricity, subject to the obvious limitation that extreme shifts in temperature are not desirable. This broader class of energy storage systems could enable significant benefits for customers that use them for time-of-use price arbitrage or distributed solar integration; they also could benefit the grid by enabling customer demand response that helps balance grid supply and demand.

APPLICATIONS OF DATA SCIENCE ELSEWHERE IN THE ENERGY SECTOR

Outside the electricity sector, data science is already transforming other aspects of how energy is produced, delivered, and consumed, and has the scope for even more far-reaching changes.

TRANSPORTATION

Both electric and internal combustion vehicles are incorporating in-vehicle sensors to monitor factors such as speed, braking, and fuel consumption. These data on performance and operator behavior could provide feedback to the driver to improve driving systems or enable energy-efficient automation strategies in the future.

A particularly exciting possibility is that of data science harnessing electric vehicles to ease some of the strain on the electricity grid from the rise of intermittent renewable generation, because EVs can double as transportation vehicles as well as mobile batteries to back up the power grid. EVs are certainly on the rise; heralded by the emergence of Tesla, traditional auto manufacturers including General Motors, Land Rover, and Volvo have signaled a permanent shift away from internal combustion engines toward EVs and hybrids. And EV fleets offer a potential grid resource when not in use and when managed in aggregate. Although companies such as Stem are building large stationary fleets of batteries to provide grid services, EVs could offer an alternative for energy or ancillary service resources. EV fleets, especially as autonomous EVs become more commonplace, could provide load balancing by moving to areas where the distribution grid is instantaneously congested. Still, this feat will require a significant coordinated data collection and optimization platform, as well as dynamic pricing signals that are temporally and geographically granular to incentivize EV owners to deploy their vehicles as mobile batteries.

OIL AND GAS

In the oil and gas sector, data analysis techniques are being increasingly used for both exploration and production. In the exploration process, voluminous amounts of seismic data are collected and analyzed to determine the presence of oil and gas. Machine learning can be used to increase the accuracy of the process to determine whether a resource worth drilling for exists. For production, diverse datasets such as well logs, imagery data, and sensor readings could be used to produce more oil and gas and do so in a way that increases safety.

ENERGY SYSTEMS PLANNING

One of the greatest energy system challenges is the overall complexity and scale of the system. The energy sector is foundational to most human activity and has aspects that are both highly centralized (traditional electricity generation) and highly distributed (vehicle and building energy consumption). Therefore, many types of information are needed for planning purposes. Some of this information is publicly available in some locations (e.g., power plant locations and capacity in the United States), others are proprietary (e.g., transmission line locations in the United States), and some others are not available without access to survey data (e.g., village electrification rates in sub-Saharan Africa). For example, by using information on transmission line locations and village electrification rate, specific villages could be identified as benefiting the most from electrification, and the optimal path toward electrification (grid connection, microgrid, or off-grid solar) could be determined.

One approach to acquiring data at this scale would be to aggregate data from diverse public sources. Remote sensing data and satellite imagery offer a supplemental data science approach to acquiring data. Previously, satellite imagery has been used to estimate the amount of oil in large oil tanks, coal consumption from changes in coal pile height, and the trade of goods by watching container ships unload their wares. These techniques rely on object detection and analysis in imagery data, and can help find power plants, estimate the energy consumption of buildings, forecast traffic flows, provide insights into electrification rates of rural villages, and map out transmission lines.[5] These analyses can be repeated whenever new imagery data becomes available, which for companies such as Planet could be daily at a global scale.

CHALLENGES

One of the greatest challenges facing most of these applications is the availability and the ease of access to data. Not all data can be made available, for several reasons, including proprietary status, being considered personally identifiable information, and logistical difficulties in the sharing process due to rigid and sometimes disparate storage systems. However, in almost every application, research and development (R&D) efforts require access to significant amounts of data to ensure the quality of the tools and insights that are produced. Establishing policies that enable streamlined data sharing for approved R&D purposes is a prerequisite for more democratized access to data. In many cases, access to data requires significant expense, which limits the number of people who can work on these problems. Increased access to open data sources coupled with deepened collaboration among policymakers, practitioners, and researchers will enable transformative innovation. Data science techniques applied to accessible energy systems data will help the work toward a cleaner, more accessible, more affordable, and more reliable future for global energy systems.

Applying Data Science to Promote Renewable Energy

Sunil Garg

The mechanical world of industrial machines and equipment is converging with the digital world because of the commoditization of sensors, affordable computing power and storage, smart software, and faster wireless connectivity. Companies in many industries, including energy, have started using new digital technologies and tools such as artificial intelligence (AI) and data science to improve their productivity, reduce costs, and increase customer satisfaction. Digitalization and AI can also speed the transition toward a cleaner electricity system in the United States and across the world by helping generate more renewable energy. But unaddressed legal, regulatory, and competitive issues could jeopardize the opportunity.

HOW ARTIFICIAL INTELLIGENCE CAN PROMOTE RENEWABLE ENERGY

Often called the industrial Internet of Things (IoT), machines today have been outfitted with dozens of sensors that produce massive amounts of data. But data alone does not create value. It needs to be combined with data science—the practice of extracting value or insight from data—to create actionable intelligence. AI is a promising new tool that can harness data to improve the operation of equipment by predicting when it will fail and enabling firms to better maintain it. Insofar as it makes renewable energy equipment cost less and produce more power, AI can make renewable energy a more attractive alternative energy source.

Sunil Garg is an executive at Uptake Technologies Inc., a Chicago-based industrial AI software company that provides predictive analysis to major industries, including those in the power sector. The views presented are those of the author and do not necessarily represent the views of Uptake Technologies Inc.

Uptake, a Chicago-based start-up at which I am an executive, ingests and normalizes data from industrial machines across a range of industries and combines that sensory data with additional contextual data such as weather. It then uses data science to generate insights and predictions that technicians act on, leading to greater productivity, reliability, safety, and security. Currently, Uptake provides this service for nearly a dozen heavy industries, including owners and operators of renewable energy resources. As a start-up, Uptake faces enormous challenges in breaking into legacy industries dominated by incumbents. Nonetheless, Uptake's AI technology in many cases has offered to incumbents benefits that outweigh the costs imposed by their preference to operate their machinery in traditional ways.

REDUCED COSTS

Even though the unsubsidized levelized cost per megawatt-hour (MWh) of electricity from wind and solar power has dropped significantly and is now lower than that from coal and nuclear, renewable sources still face competitive pressures.[1] For example, some existing wind parks are producing less energy than preconstruction projections and will generate less revenue when the federal production tax credit for wind power output phases out at the end of 2019.[2] For projects still under development, developers should achieve an extremely low cost per MWh of power produced by the finished plant, because potential buyers of the electricity are seeking to sign extremely low-cost contracts at energy procurement auctions.

Since March 2017, Uptake has been providing predictive insights to MidAmerican Energy for its wind turbine fleets, which total approximately five gigawatts (GW) of capacity. During the first few days of monitoring one of the sites in early 2017, Uptake's software spotted anomalies and predicted the future failure of one turbine's main bearing. The main bearing is a critical component of a wind turbine; it helps hold up the rotating parts of the turbine and transmits structural loading to the tower. A technician indeed found problems with the main bearing, and a few hours of preventive maintenance was performed that cost approximately $5,000. That work helped avoid a surprise failure that would have resulted in several days of downtime, lost power production, and a costly repair, all totaling up to $250,000 in lost value.

Additionally, Uptake's software incorporates weather predictions so that required maintenance can be performed when wind speeds are too low to generate energy. The company plans to incorporate energy market forecasts into its software to provide insights to operators so that their turbines are ready to produce power when the price for energy is high.

INCREASED RELIABILITY

The ability to preemptively fix problems increases operators' confidence in the amount and timing of energy production, enabling them to sell more confidently in the day-ahead energy markets. It also allows utilities and grid operators to rely more on renewable energy generation and less on fossil fuel generation (coal, natural gas, and oil). In the past decade, almost every power plant to have been retired in the United States was powered by fossil fuels, and in 2018 an estimated 13 GW worth of coal-fired plants are scheduled to retire.[3] For this trend to continue as renewable energy occupies a more material share of electricity than in the past, grid operators will need substantially more confidence in their projections of future renewable output in the absence of highly flexible conventional energy sources.

Uptake's AI has helped improve the reliability of wind turbines and thus enabled grid operators to regard wind turbines as a resource that they can count on to produce power. In one instance, Uptake found a wind turbine reporting no wind although it was near another turbine that was producing energy. Additional investigation revealed that the non-energy-producing turbine's anemometer, the instrument used to measure wind, was miscalibrated. This error had prevented the control system from permitting the blades to turn and produce energy. By identifying the error, Uptake enabled the wind turbine to produce power when the grid operator's weather forecasts would have predicted the turbine should be producing power, enabling the operator to count on the wind turbine as a dependable, rather than capricious, resource.

INCREASED RENEWABLE ENERGY PRODUCTION

Downtime, both planned and unplanned, is a big challenge for the industrial sector. It is also the most expensive and is especially costly for intermittent energy sources such as wind and solar. Each year, industries across the globe lose \$647 billion due to downtime.[4] Eliminating downtime increases overall renewable energy production. A January 2018 report estimates that downtime in the current U.S. wind fleet prevents at least 12 terawatt hours of energy from being generated—enough energy to power 1.1 million homes, the equivalent of all the homes in Chicago, for a year. Producing this renewable energy would also reduce carbon dioxide emissions by 8.9 million metric tons per year.[5]

AI can be used to increase efficiency in thermal energy from conventional power plants as well—generating more energy from a given amount of fuel—and thus reducing overall environmental damage. A 2010 study by the U.S. Department of Energy's National Energy Technology Laboratory found that increasing the efficiency of U.S. coal plants from 32 percent to 36 percent would help reduce 175 million metric tons of greenhouse gas emissions per year.[6]

HOW POLICY CHANGES CAN ADDRESS ADOPTION CHALLENGES

Despite the promise of and early results from new digital tools such as AI and data science, several issues currently prevent wider adoption and greater generation of renewable energy generation. Uptake would better be able to serve its customers and promote renewable energy if policymakers pursued three initiatives.

ALIGNING INCENTIVES FOR UTILITIES AND CUSTOMERS

Three Supreme Court cases from the twentieth century provide much of the modern-day framework for balancing the interests of investors in and consumers of utilities.[7] While utilities are allowed to recover their operating costs, regulators have interpreted these court opinions to allow for a "just and reasonable" rate of return on only capital.[8] Although well intentioned, this construal, which allows for only capital expenditures to receive a rate of return and narrowly defines capital, disincentivizes investing in new technologies that are not hardware focused. Computer hardware, such as servers and computers on-site, are generally counted as capital, but new digital tools that come in the form of software service or via the cloud are not.

Lately, leaders at the federal and state levels have been paying attention to this challenge. On November 16, 2016, the National Association of Regulatory Utility Commissioners adopted a nonbinding resolution that calls for regulators to make "software procurement decisions regardless of the delivery method" and encourages "state regulators to consider whether cloud computing . . . would be eligible to earn a rate of return and would be paid for out of a utility's capital budget."[9]

Proposals in the U.S. Congress and in several states seek to promote digital tools that lead to greater renewable energy generation. Senator Angus King has proposed legislation that would require state regulatory authorities to consider broadening the scope of what expenses can be factored into rates, including performance-based incentive mechanisms that lead to environmental and other benefits.[10]

Some states have also started to consider proposals or solicit feedback on modifying what is allowed to be included as part of their rate base, a measure of the capital investments that a utility's shareholders have made and on which they can secure a rate of return by recovering customer rates. In 2017, a report issued by the Illinois Commerce Commission found that current rules create "a disincentive for utilities to invest in new technology" and that "a strong consensus that current regulatory accounting rules have not kept pace with technological innovation" exists.[11] In New York, as part of its Reforming the Energy Vision effort, the New York Public Service Commission ruled in May 2016 that utilities can include the lease costs of software in their rate base.[12]

Wider adoption of these early proposals, along with new efforts, can remove barriers currently slowing the adoption of digital tools that can increase the prevalence of renewable energy. Congress should also pass legislation that encourages and incentivizes regulatory bodies to allow experimentation with digital tools that help create a cleaner grid.

RESOLVING DATA OWNERSHIP AND ACCESS DISPUTES

In recent decades, manufacturers of industrial machinery have tacked on maintenance service contracts along with sales of their machines to increase revenue and profit. Many machines, from construction tractors to wind turbines, now have sensors and onboard computers that operate with proprietary software and generate data. Disputes have arisen, however, regarding the ownership of and access to this data and whether restrictions can be placed on machine owners and third parties.

In 1998, Congress passed the Digital Millennium Copyright Act to update U.S. copyright laws for the digital age and to provide a framework for protecting copyrights in software. The law prohibited users from circumventing technical measures to protect copyrighted materials. However, it also allowed exemptions to this prohibition to be granted every three years for cases in which users would be "adversely affected" by the law's prohibition on circumventing technical measures.[13] In 2015, the U.S. Copyright Office adopted an exemption that permits owners of land vehicles (such as tractors) to access data and electronic control units to the extent necessary to allow the diagnosis, repair, or modification of their vehicles.[14] But the decision has only deepened the dispute, and the issue remains unresolved partly because of the narrowness of the exemption and the fact that it does not permit third-party access to the data.

The triennial rule-making proceeding for regulators to consider whether additional or wider exemptions are warranted is currently under way. Important arguments on both sides of this issue exist. Manufacturers of industrial machinery often want to restrict access by customers and third parties to the data generated by these machines because the manufacturers believe they are best equipped to perform maintenance, that permitting customers and third parties to perform maintenance could open thorny liability issues, and because manufacturers have an exclusive right to the data by virtue of having manufactured the machine. Uptake, by contrast, believes that equipment owners will be best served if a flourishing range of third-party firms—both start-ups and large companies—can compete for the business of maintaining customer machinery at the lowest cost and the highest efficacy as possible. For the latter scenario to be fully realized, however, open access to data generated by any machine or equipment for third parties and owners is critical.

ADDRESSING CYBERSECURITY THREATS

Throughout history, technological and scientific progress have ushered in new opportunities. By the mid- to late-twentieth century, advances in nuclear technology led to the generation of abundant electricity. However, as the partial meltdown of the Three Mile Island nuclear facility near Middletown, Pennsylvania, in 1979 demonstrates, these opportunities can come with significant risks too. Regulators mandated many changes, and the nuclear industry (following the Kemeny Commission

Report's recommendation) established the Institute of Nuclear Power Operations (INPO). This organization has helped prevent another major meltdown from happening in the United States. The INPO helps promote nuclear safety and reliability through its facilitation of training, accreditation, information exchange, and technical assistance.[15]

As is true of nuclear power, IoT provides the opportunity to both generate more renewable energy and create a more reliable and secure world. However, machines and devices are becoming connected faster than they can be secured from cyberattacks. Cybersecurity breaches into information technology (IT) devices such as consumer devices are costly, but breaches of a growing subclass of IT hardware and software known as operation technology related to industrial equipment risk human life and national security. Most consumer software is widely tested and constantly updated, but the number of unique protocols in industrial software leaves it vulnerable to cyberattacks.

Leaders and companies in the industrial IoT should form a governing structure that proactively sets standards, provides technical assistance and guidance, and establishes best practices to prevent catastrophic cyberattacks. Much is at stake. These are issues of consequence and not simply convenience. Too many people depend on the machines that power homes, transport people, and make the modern world work.

Part III: Managing the Risks of Digital Innovations

Managing the Cybersecurity Risks of an Increasingly Digital Power System

Erfan Ibrahim

Before the advent of the internet, electric power systems were either not digitally networked (and thus relied on proprietary protocols owned by single organizations) or connected in the simplest of configurations. Unless a user was directly connected to the legacy electricity system through a dedicated terminal, it was difficult to disrupt the application running on the system. The lack of networking or limited knowledge of proprietary protocols made it nearly impossible for a remote hacker to connect to the legacy system, let alone disrupt it logically.

This security-by-obscurity model of legacy data communication technologies started to fall apart as Transmission Control Protocol/Internet Protocol (TCP/IP) became popular with the growing public interest in the internet. Although TCP/IP was developed in the 1960s, its use was limited to government and academia until the late 1980s. In 1990, the TCP/IP-based U.S. Advanced Research Projects Agency network (ARPANET), the National Science Foundation network (NSFNET), and the Defense Advanced Research Project Agency network (DARPA-NET) merged to form the backbone of the public internet.

At that time, TCP/IP became routable, allowing users to link to information systems remotely without a dedicated link between source and destination. An increasing number of economic sectors began to see the value of the TCP/IP to carry data traffic between two or more users and systems to support critical business functions. The electric power sector also started to leverage TCP/IP for its information technology (IT) and operational technology (OT) data communication needs. The economic boom of the 1990s was largely based on the development of this communication and IT infrastructure. The business world started to migrate from analog- and proprietary-based digital communication systems to TCP/IP open standards-based systems for operational transactions.

The rise of TCP/IP led to the rapid deployment of low-cost, networked distributed energy resources across the globe. Consumers of electricity turned into prosumers (producers and consumers) of electricity. Deregulation of the global electric sector allowed independent power producers (IPPs) to generate power and offer it in the energy markets by receiving pricing signals and sending telemetry data over TCP/IP-based networks to the power balancing authority. This put IPPs in competition with the regulated utility monopolies, in many cases lowering the cost of electricity to the end user.

The upsides of this digitalization and interconnection in the energy sector are apparent from economic, technological, and efficiency perspectives. Those benefits also come with drawbacks, however, such as cyber vulnerabilities, which threaten grid reliability and national security. A network that links a legitimate remote user to a system can also link a remote hacker to the same system, potentially leading to disruption of the IT systems, sensitive data exfiltration, or proliferation of malware across multiple interconnected networks that can cause large-scale electricity outages. Additionally, legitimate users—with all the credentials to access trusted systems—could become disgruntled and act as insider cyber threats. Finally, remote hackers could use sophisticated social engineering or phishing schemes to lure an unsuspecting user to provide access to sensitive data in IT or network management systems through emails containing nefarious web links and malicious code.

Cyber threat vectors are evolving even faster than the digitalization of the energy sector. It is becoming difficult for power plant owners, for example, to embrace the high level of digitalization and interconnectedness that the wholesale power market requires of them to stay competitive, while maintaining a healthy cybersecurity posture against insider and external cyber threats. The exponential rise in the number of networked-nodes in the power grid—each with its own firmware vulnerabilities—makes it hard to monitor the entire network for cyberattacks.

The old cybersecurity architecture model was largely based on firewalls, antivirus servers, and intrusion detection software, and on providing free rein within the trusted network to an authenticated user. This model is no longer effective against insider threats or against external threats that find a way into the trusted network through a backdoor. A new cybersecurity model is needed to protect against today's fast evolving cyber threats.

CYBER GOVERNANCE AS THE FIRST LINE OF DEFENSE FOR ELECTRIC POWER SECTOR ENTERPRISES

Electricity system asset owners, such as utilities, that rely on today's digital communication technology to support their business functions need to develop and implement a comprehensive cyber governance regime in their organizations to avoid common cyber breaches from negligence, mismanagement, or lack of training. The U.S. Department of Energy (DOE) and the National Institute of Standards and Technology (NIST) have developed cyber governance standards to help both enterprises in the energy sector and other industry verticals (i.e., transportation, health care, and government) enforce a robust cyber governance regime within their business units.

The DOE cyber governance standard is based on its Cybersecurity Capability Maturity Model (DOE C2M2) and NIST's standard is based on its Cybersecurity Framework (NIST CSF). DOE C2M2 has 386 business process security controls defined across ten domains: risk management, asset change and configuration management, identity and access management, threat and vulnerability management, situational awareness, information communication and sharing, event and incident response, external dependency management, workforce management, and cybersecurity program management. The DOE standard provides a way of assessing an organization's cybersecurity controls through four maturity indicator level (MIL) designations: MIL 0—no process; MIL—ad hoc process; MIL 2—managed process; and MIL 3—adaptive process. An organization should strive for the highest MIL levels across its various cybersecurity controls.

Electric power utilities should undertake routine cyber governance assessments to ensure advanced levels of cyber maturity in handling digital technologies for a variety of businesses and organizations. The results of such assessments can provide prioritized action items to each utility so that both management and cybersecurity staff know the order in which security controls from the integrated model should be implemented. The weight of each prioritized action item should depend on the centrality of the control (how many other controls depend on it) and the maturity level of the control (the lower the maturity level, the higher the priority for implementation).

DEFENSE-IN-DEPTH CYBERSECURITY ARCHITECTURE FOR DIGITAL NETWORKS

In addition, utilities should pursue a defense-in-depth strategy to improving cybersecurity that creates multiple overlapping layers of protection. This paradigm is distinct from, and superior to, a strategy of simply fortifying the boundary around the electric power system and simply trying to keep intruders out. Instead, a cybersecurity architecture that strives for defense-in-depth consists of four functional layers.

The first layer of defense is a set of controls that ensure that only users with a legitimate organizational role and a bona fide need to access particular data or equipment are granted that access. This layer sets confidentiality requirements for the network that are enforced by authentication, role-based access control, and cryptography. All other data is either inaccessible by privilege rules or encrypted so that it cannot be used without formal authorization. Examples of techniques to enforce such controls include firewalls with remote logins via virtual private networks between remote users and corporate firewalls, blocking or selective encryption of data in the enterprise, access control lists on switches limiting data traffic between network nodes, username/password locks on IT or OT applications, and digital certificates for authentication.

Still, malicious actors could seek to breach the first layer of defense, at which point they will trigger the second layer. This layer seeks to detect malware signatures that have intruded into the network and take measures to restore network security. Such a second layer is common on the IT side of corporate networks and increasingly common on OT networks that are moving to TCP/IP and prone to malware originating from the public internet. Many products are commercially available in this space to implement a second layer of defense. For example, intrusion detection systems are available from Cisco, Juniper, NexDefense, N-Dimension, FireEye and Palo Alto Networks. NexDefense's Integrity product looks for anomalies in network connections and ports serving specific applications from certain IP nodes and alarms in the event of violations relative to expected network connections and protocol traffic between IP nodes. Products such as N-Dimension N-Sentinel go a step further by customizing their offerings for the electric power sector and detecting signatures for malware from power systems' supervisory control and data acquisition SCADA protocols such as

DNP3, IEC 61850, Modbus, and OPC. Palo Alto Networks and FireEye use pattern recognition of cyberattacks to develop near zero-day attack protection (i.e., protection against an attack on a vulnerability that was previously unknown) for IT type malware proliferation. The second layer of defense is enough to provide situational awareness for IT security systems and to protect against a limited set of nefarious behaviors on OT networks that leverage internet-based malware (an example is the breach that led to the Black Energy attack in Ukraine that knocked out large parts of the power system).

The first and second layers of security alone might fail to protect against insiders with network access credentials or an outsider who steals access control credentials to the power system but lays dormant for long periods. The latter is known as an advanced persistent threat and might perform nefarious activities on specific devices in the power system about which the actor has deep domain expertise and can disrupt systems or exfiltrate data with great subtlety. Because this type of hack does not come with any malware that has a signature, the tools of the second layer of defense, which might be enough to safeguard IT systems but not OT ones, cannot catch it.

To guard against such threats, the third layer of security sets up context-based intrusion detection and prevention capabilities. These capabilities are achieved by setting up intrusion detection systems that can sit on inline taps (a method of monitoring a network) or hardware layer filters, and allow certain commands and values in a messaging protocol to be transmitted over the network but block others. Such systems understand the semantics and commands of the protocols that they are observing and check the messages and values relative to a set of rules set up by the operator. Firms that offer products that can implement the third layer of defense include Albeado, SUBNET, SecLab, CyberX, WaterFall Security, Indegy, and Radiflow.

Finally, the fourth layer of security is endpoint security, which protects against breaches targeting the equipment at the edges of the electric power network, whether customer appliances or electric grid distribution grid equipment. Implementations of this layer include firewalled virtual machines known as hypervisors, firewalls at the operating system level either directly installed on hardware or a virtual machine instance with an operating system, username and password protection on a software application, encrypted data stored on the end system with select keys for certain data fields, and tamper-resistant software that alerts the system operator if the host application is compromised and auto-ejects from the network so it does not become a staging ground for a wider cyberattack.

Endpoint security in power systems can go further into the realm of non-networked security. At the microprocessor level, the power system node is programmed to accept only certain commands in the SCADA protocol based on its function and to reject others. If the state of the system deteriorates, a self-restorative force in the endpoint enables a return to an acceptable range of values for variables like electric voltage, current, and phase. This type of security is not networked because this logic is programmed into the microprocessor and cannot be altered remotely. This precaution is critical to protecting high value power systems assets that could be targeted by insiders or advanced persistent threats.

PROTECTION AGAINST DATA PRIVACY BREACHES

The three main sources of customer data privacy vulnerabilities in energy systems exist at the source of the generated data (e.g., smart meters), during transmission of the data (e.g., from smart meters to utilities or third-party aggregator networks), and in data storage (at the utility or third-party aggregator meter data management system). Insider cyber threats at utilities, third-party aggregators, or external cyber threats using sophisticated social engineering skills could get access to customer data and steal, modify, or disrupt it for a variety of reasons. The result can be a public relations catastrophe in addition to possible legal and financial consequences for the energy systems asset owner.

Confidentiality of data in smart meters and during transmission can be protected with advanced encryption techniques on the smart meter and the advanced metering infrastructure network, from the smart meter to the meter data management system at the utility or third-party aggregator. The downside is that symmetric encryption key management over large geographically distributed areas is a challenge—replacing a compromised encryption key quickly and effectively is difficult. Encryption is often too resource intensive on the digital systems that are required to support it (i.e., memory, processing, and networking burden of encryption).

If a power system asset owner carries out cyber governance assessments of their critical business units and implements the four functional layers of the cybersecurity architecture, they can mitigate the risk of data privacy breaches and protect customer data against theft, tampering, or misuse from insider and external cyber threats.

Managing the Economic and Privacy Risks Arising From Digital Innovations in Energy

Jesse Scott

The growing use of digital technologies in the energy sector raises questions about data protection, in particular personal data privacy and ownership, and also threatens to upend job markets.

Smart grids and digital demand response technologies rely on utilities or aggregators collecting vast quantities of consumer-specific, real-time energy use data. In the case of active management of behind-the-meter appliances in a smart home, this data includes records of personal energy use events, such as heating water for a shower or opening a refrigerator—giving anyone who has access to the data a snapshot of householders' daily routines and activities. How much information people are comfortable sharing with service providers, how confidentiality can best be protected, who owns this type of consumer-specific data (the individual who generates it or the service provider who collects it), and who can use or share data and for what purposes are becoming critical questions.

Digitalization is changing work patterns and tasks, causing job losses in some parts of the energy sector and creating new jobs in others. Smart meters and coders are supplanting manual meter readers. Workers supporting digital infrastructure need specialized information and communication technology (ICT) skills, including in cybersecurity, and many other workers now need basic ICT skills to operate digital technologies. However, digitalization is unlikely to replace the sizable labor force needed for major engineering and construction activity related to physical infrastructure such as grids and pipelines.

Still, despite these risks, digitalization also has the potential to bring enormous benefits, both for market liberalization and decarbonization. The November 2017 International Energy Agency (IEA) report on digitalization and energy, of which I was a coauthor and from which this essay draws insights, therefore, urges policymakers around the world to collaborate across borders and with the private sector to manage the risks of digital innovations while reaping their benefits.[1]

RISKS TO DATA PRIVACY AND DATA PROTECTION AND OPTIONS TO MITIGATE THEM

Privacy poses challenges for the energy sector. Regarding data from smart electricity meters, consumers who wish to protect personal anonymity are aware that detailed behind-the-meter energy data could be used to find out when buildings are empty. For industrial and commercial consumers, demand response infrastructure that tracks energy use could reveal information about proprietary business practices and operations. Data breaches are a central issue affecting customer willingness to use new services.

At the same time, consumers and companies will sometimes want to share their energy data with third parties—such as in a marketplace for energy efficiency services. Providers also see opportunities to monetize energy use datasets, which could be a new revenue source. Utilities in New York State have proposed that regulators allow utilities to charge a fee for making granular, value-added data (such as aggregated customer usage information) available to local distributed energy resource providers.[2]

Energy companies are increasingly subject to stronger data protection regimes. The European Union's General Data Protection Regulation (GDPR), which came into force on May 25, 2018, introduces requirements for privacy and customer consent for data collection and use to be designed into all business processes for products and services. It also establishes a right to data portability, the transfer of personal data from one service provider to another. GDPR requires significant changes to business practices for many companies, including foreign-based companies that collect or process the data of EU residents. Businesses will need to compile data records, understand what data have been stored, and how different departments are using customer data. They will also need to maintain a strong audit trail of permissions that customers have given for use of their data and how this information flows through an organization. Some businesses will need to hire a data protection officer. Meeting the standard will entail significant investment to implement process, personnel, and infrastructure changes.[3]

Elsewhere in the world, many older legal frameworks regarding privacy, consumer protection, and electronic communications do not adequately establish up-to-date technical definitions. For example, in the United States, whether transmission of consumer energy usage data over the smart grid is an electronic communication under federal law is unclear.[4]

Lack of clarity or robust protections could discourage consumers from participating in sharing digital energy information and therefore limit the potential for overall system efficiency and possibly lead to a consumer backlash. Viewed from an overall energy systems perspective, however, a strong public interest case exists for making aggregate data widely available. For instance, large datasets can provide insights for urban planners or allow researchers to investigate aggregate efficiency opportunities that cannot be realized by individual consumers on their own. Privacy concerns need to be balanced with promotion of market innovation, the operational needs of utilities, and the wide-ranging potential of the digital transformation of electricity.

Companies and policymakers can use technical tricks to balance privacy and innovation. Companies can make data anonymous by aggregation so that private information cannot be attributed to a specific household. Another approach is to limit data granularity by adding a time lag, which makes it more difficult to track individual energy use events. In Germany, the law on smart meter data allows transmission of household data only every fifteen minutes; in France, ten minutes is considered enough for smart grid operations.[5]

Policymakers also need to consider whether regulation should take an opt-in or an opt-out approach to customer authorization. Opt-in programs offer customers maximum protection and require affirmative customer authorization for certain data to be shared. Opt-out programs, however, are more likely to favor mass participation in demand response markets. Alternatively, customers could be given a range of confidentiality options. "Minimum" customers could choose to participate only in enough data collection to enable core smart grid operations, such as load balancing and price formation. "Maximum" customers could agree to detailed data being made available for marketing purposes to commercial energy efficiency providers, the aim being to learn about possible savings. One example of an approach that offers both opt-out and opt-in components is a voluntary code of conduct for utilities developed by the U.S. Department of Energy and the Federal Smart Grid Task Force in 2015 that distinguishes between using opt-out schemes to collect primary data that can be "reasonably expected" and are crucial to more efficient grid operation, and requiring opt-in consent to collect a more extensive set of secondary data.[6]

Despite the risks, managing privacy and data protection has an upside. When companies and regulators approach the challenge intelligently and proactively, robust processes and systems can help strengthen customer communication. The UK information commissioner emphasizes the opportunity for first mover companies to establish high trust and confidence in their digital products and services.[7]

Privacy is culturally specific, and each country faces different opportunities, priorities, cultural contexts, and circumstances. Few one-size-fits-all answers exist. Consider three diverse jurisdictions: California, France, and South Korea. In 2016, French legislation established a broad concept of public interest data, enabling the government to request data from commercial entities to establish public statistics.[8] Since 2011, California and South Korea have led the world in developing strong new privacy protections specifically for smart meter data. California requires utilities to encrypt usage data. These provisions have also been extended to internet service providers, financial institutions, and other businesses that could handle or receive smart meter data.[9] South Korea has focused on setting robust limits on data collection, use, outsourcing, disclosure, editing, searching, storage, and destruction.[10]

RISKS TO JOBS AND SKILLS

In some ways, digital technologies are causing companies to retrain or replace parts of their workforce.[11] For example, in the life cycle of a power plant, digitalization has the greatest effect on equipment, including its manufacturing, siting, and operation and maintenance; by raising the productivity and reliability of the plant, digitalization therefore potentially reduces labor intensity. In thermal electricity generation, digitalization could change the tasks of existing operation and maintenance of power plants, while creating new jobs in data science. In the renewables sector, robots can be used to clean solar panels and drones to monitor wind turbines.

A large share of employment in upstream oil and gas is associated with initial field development. Digitalization and other innovations have helped lower costs and raise productivity, although reductions in employment are difficult to disaggregate from the wider effects of the lower oil price environment. The widespread use of 3D and 4D seismic analysis has reduced drilling needs but created new jobs in ICT that require different skill sets and are often located in a different region from drilling operations.

In the energy efficiency space, most jobs are involved with the initial labor (e.g., construction and refurbishment of equipment for building retrofits), and fewer jobs are associated with the operation of the equipment once installed. Installers and technicians working on the operations and maintenance of buildings, nonetheless, will likely need additional skills to deal with new technologies. Skills training could become a barrier to energy efficiency.

Significant attention has been focused on the potential job losses autonomous cars and trucks could lead to. The United States alone has around 3.5 million truck drivers, 665,000 bus drivers, 230,000 taxi drivers, and at least 500,000 active drivers with the ride-sharing companies Uber and Lyft. If technological change and adoption are rapid, digitalization could transform the transportation sector. However, public acceptance and the regulatory environment are as important as technology in determining the pace of deployment of autonomous vehicles, and because uncertainty over all three variables is substantial, the pace at which job displacement will occur is unclear. In fact, at the moment, the opposite of job displacement—increased hiring—appears likely, given that the International Transport Forum reports that the U.S. trucking industry already has a shortage of around fifty thousand drivers, which will grow to eight hundred thousand by 2030.[12] Moderate to high levels of automation in trucking, especially in the coming years, could be deployed to complement human drivers and help fill the shortage of available drivers, which would not cut into existing jobs but rather displace growth in jobs.

POLICY RECOMMENDATIONS

Policymakers should work with energy industry experts, colleagues across departments, and their counterparts in other countries so that they can understand and better manage the risks of economic displacement and privacy breaches arising from digital innovations in energy.

Participate in interagency discussions. Managing the risks of digital innovation requires energy companies and energy policymakers to collaborate beyond the usual boundaries of the sector. Many jurisdictions around the world are developing digital strategies for their economies. For example, since May 2015, the European Commission has delivered thirty-five legislative proposals and policy initiatives in its Digital Single Market Strategy.[13] Policymakers in the energy field need to participate fully in government-wide decision-making about digital regulations and should actively work with colleagues across their governments to track the implications of digitalization and digital regulation for energy operations and business.

Build expertise. Energy policymakers need to remain well informed about the latest developments in the digital world: its nomenclature, trends, and ability to revolutionize a variety of energy systems (in both the short and the long terms). A major part of this endeavor consists of ensuring that energy policymakers have access to staff with digital expertise. Education policies and technical training to ensure an adequate pool of relevant expertise for both the private and public sectors are critical.

Learn from others. Each country is different in ways that are relevant to digitalization's increasing effect on energy systems; nonetheless, lessons can be learned from the experiences of other regions and jurisdictions. These lessons can include both successful case studies as well as cautionary tales. Useful collaborations and best policy-sharing can take place in a variety of forums, including a wide range of IEA Technology Collaboration Programmes.[14]

Part IV: Policy Recommendations

How State-Level Regulatory Reform Can Enable the Digital Grid of the Future

Richard Kauffman and John O'Leary

Nikola Tesla and George Westinghouse would immediately recognize the fundamental architecture of today's electric grid, which remains largely unchanged from its earliest days. Similarly, the policy and regulatory environment that surrounds the physical system is much the same today as it was at the outset. This system was not intended to be energy efficient, nor was it designed for distributed energy resources or intermittent renewable power. It is, in many respects, an analog system. But the power grid of the future will embrace the value of digitalization: it will permit dynamic balancing of electricity supply and demand and more activity by prosumers (consumers who also produce energy), and it will easily accommodate the intermittency of renewable energy.

Although technological challenges will need to be overcome to build this digital grid, the regulatory and policy regime developed alongside the original grid architecture poses even larger challenges. In the same way that wind and solar power generation and storage have been attached to a physical system never designed for those resources, new policies are being incorporated into a regulatory regime intended to build yesterday's grid. A new grid needs a fundamental change in the policies—most of which are decided at the state level in the United States. The good news is that lessons from other sectors of the economy can be applied to drive investor capital to build the new grid.

Other capital-intensive industries have faced unrelenting global competition over the past four decades, especially as many have been deregulated. This competition has compelled capital-intensive industries—from airlines to automobiles and chemicals—to change by adopting new technology and embracing new business models. These changes have markedly improved capital efficiency, created savings, and increased value to customers.

For example, on the eve of deregulation, the U.S. airline industry had an average capacity utilization in the mid-50 percent range; thanks to the adoption of dynamic pricing, technology-enabled route optimization, and other changes, the industry's capacity utilization has climbed past 80 percent.[1] This improvement has meant much cheaper transportation costs, and the industry has never been safer. Similarly, the telecom sector demonstrates how an entirely new network can be built in little more than a decade. The U.S. government did not determine the number of modems or mandate a certain number of new switches. Instead, deregulation allowed market forces to allocate capital and promote innovation.

Meanwhile, the regulated power sector has remained mostly untouched by the forces that have driven innovation in the rest of the economy. The sector's capacity utilization remains at about 50 percent, and customers pay all year long for a system that has enough power generation and delivery capacity built in to meet the highest few hours of customer demand in a year.

Although the regulated power sector will likely remain a natural monopoly for some time, fundamental change in state-level regulation to build the new grid requires stripping away regulation that is not required today while opening the sector to forces that already prevail in other sectors of the economy. If this happens, the electricity system could transform dramatically, and digital innovations will be at the center of that transformation that enables a cleaner, cheaper, and more reliable and resilient power system.

REMAKING UTILITY FINANCIAL INCENTIVES

A regulatory construct known as the rate base underpins the utility business model across the fifty U.S. states and is a barrier to investment in innovation, digital or otherwise. The rate base is a measure of the capital that a privately owned utility has invested in infrastructure, and to enable utility shareholders to recoup those investments, state regulators authorize utilities to collect rates from customers that cover the costs of paying off the infrastructure and also pay utility shareholders a rate of return to compensate them for investing their capital in the electric power system. Because customer rates cover the electric system's costs plus a shareholder rate of return on infrastructure, this regulatory model is sometimes called the cost-plus model.

Under the cost-plus model, utilities have a direct disincentive to support digital innovation or even upgrade their internal information technology (IT) infrastructure. The issue is that state regulators do not consider IT investments as traditional infrastructure investments, and therefore an IT investment would not increase the size of the rate base, on which shareholders receive returns; therefore, utility executives are less likely to invest in digital infrastructure for fear of reducing their shareholders' return on equity. More insidiously, if strong IT capabilities were to improve utility system operations and, by extension, identify ways to reduce capital expenditures (e.g., by avoiding investments in new poles and wires), then that efficient IT investment would shrink the utility's revenues and profitability under the rate base construct.

It gets worse. Hundreds of companies across dozens of industries cannot currently take advantage of the utility platform—both the physical electricity grid and the as-yet unbuilt IT system—to drive efficiency, improve operations, or provide more value to customers. The opportunities range from developing self-sufficient energy microgrids that network distributed energy resources, to intelligently deploying electric vehicles to help balance grid supply and demand, to enabling peer-to-peer electricity trading facilitated by blockchain-based ledger accounting. All that third-party potential, which could decrease the overall cost and improve the value of the electricity network, is currently locked out of the system because utilities have no incentive to let them in.

These promising approaches and many others, putting to work many technologies that exist today, can gain scale only if the utility provides a platform capable of orchestrating third-party technologies and business models. And those capabilities will come into existence only if the utility has a financial incentive to provide them. Reformed utility financial incentives can create a pathway to innovation and deployment of the associated digital technologies that could yield enormous value to the energy system and its customers.

TAKING A NON-WIRES APPROACH

The state of New York has sought to transform its electricity system and regulatory bodies into a laboratory of policy, economic, and technical innovation. The campaign to reimagine the energy system is called Reforming the Energy Vision (REV). One of the most promising and scalable examples to come out of REV is the non-wires approach. When confronted with a system "problem"—the growth of customer electricity demand in a particular area of the network, or the need to replace or upgrade capital-intensive infrastructure—instead of defaulting to a poles-and-wires rate base investment, New York utilities now have an incentive to work with the market in pursuit of cheaper, cleaner alternatives—and in most cases, those alternatives will take advantage of digital innovations.

To make this possible, New York State has altered utility regulations so that across a variety of initiatives, utilities can earn a profit by retaining a share of savings created for ratepayers. That savings can come from a wide range of activities, whether implementing a new IT system, deploying an intelligently operated battery project, or many other digital or physical technology approaches. Utilities are encouraged to make public requests for solutions from the private sector and evaluate the proposals based upon the savings they generate for customers—without compromising goals such as electricity reliability—because customer savings translate directly to utility profits.

The Brooklyn-Queens Demand Management program is an illustrative example. In that program, the utility Con Edison solicited from the competitive market a $200 million portfolio of distributed energy resources to defer the construction of a $1.2 billion substation. This is no isolated example, however. Utilities across New York State have adopted the non-wires approach, and thirty-six non-wires projects in various stages are moving forward. These projects will yield billions of dollars in ratepayer savings, drive hundreds of millions of dollars of investment in distributed solutions, and save millions of tons of greenhouse gas emissions (the four most advanced projects alone will avoid 875,000 tons of carbon dioxide). Rolled out across the United States, the non-wires approach could yield billions of tons of carbon savings.

Migrating away from the rate base approach to shared savings compensation is only part of the answer. Processes need to change as well, not only at the companies but also among regulators. Because utilities are concerned about regulators permitting the recovery of their costs through customer rates, they use requests for proposals (RFPs) for most procurements. Although this approach makes sense for purchasing commodity equipment, it deters innovative companies from partnering with utilities: a company would not put forth its best ideas only to have them shopped to the lowest bidder.

The Con Edison Brooklyn-Queens examples demonstrates the benefit of open-ended request for solutions (RFS) approaches. This approach

could be expanded beyond capital planning to other areas of grid operations, including seeking solutions to balancing load, integrating distributed energy solutions, or using demand side resources to balance
intermittency of renewable resources. Rather than a technologically
prescriptive RFP seeking the lowest bid on a predetermined approach,
an RFS allows the utility planning process to benefit from open-source,
market-driven innovation. An RFS makes available to the private sector
a wider range of utility system data, and based upon that broader set
of information, third-party companies can propose solutions that may
contain innovative ideas otherwise not considered by the utility.

To be clear, the broader use of these processes would require meaningful change at the state-level regulator as well. The regulator of the
future would need to evolve from judicial decision-making to market-
monitoring. As a practical matter given pace of technology change,
without change in regulatory models and processes, regulators will find
themselves increasingly reactive.

Again, because the current utility business model is both energy
inefficient and financially inefficient, the new grid can be built and
utilities can be offered attractive returns without increasing customers'
bills. The more savings the utility can deliver, the greater its share of the
profit will be; this possibility could finally realign incentives to motivate
innovation that benefits utilities and all customers.

PROVIDING POLITICAL OXYGEN

Migrating incentives in any large institution is difficult, but the energy
sector presents unique challenges. The power system touches everyone—including ratepayers but also more highly organized and vocal
groups such as the business community, incumbent generators, utilities, and environmental advocates.

One reason for the progress made in New York State under REV
is that all involved groups were dissatisfied by the old system. Despite
supply costs being historically low, customer bills were still climbing in
response to increasing transmission and distribution costs; environmentalists and clean energy developers wanted more renewable energy more
quickly; and fossil fuel plant operators were dissatisfied with short-term
capacity payments supporting much of the generation fleet. In an environment in which all parties are moderately dissatisfied, elected officials

can easily enact a policy intervention that costs no money, faces minimal political opposition, and has quick and visible results.

Policymakers should collaborate with and narrow the gap between the parties, particularly given that this issue of power sector regulatory reform enjoys broad popular support, creates economic development, fights climate change, and serves the public good. Reforming power sector regulatory constructs and processes to drive both capital and digital innovation to build the grid of the future could do all that.

Lessons From Singapore's Approach to Developing Clean and Digital Energy Systems

Hiang Kwee Ho

Many analysts have characterized the energy transformation taking place across the world in terms of three shifts: digitalization, decentralization, and decarbonization. The first—digitalization—is the linchpin that enables the other two, and in the process makes it possible for countries to achieve multiple policy objectives concurrently. This is a break from the conventional calculus of energy policymakers, who have traditionally found trade-offs when seeking to achieve energy security, economic competitiveness, and environmental sustainability. Often, they give energy security and economic competitiveness precedence over environmental sustainability, to the detriment of clean energy deployment.

But the energy world is changing rapidly, and Singapore is also reviewing its long-term clean energy aspirations and efforts. Singapore plans to centrally integrate digitalization into its plan to further reduce the carbon footprint of its energy system while maintaining high levels of energy security and economic competitiveness. Doing so will require speeding digital innovations, and Singapore plans to be an international leader in fostering such innovations.

EXAMINING SINGAPORE'S CLEAN ENERGY GOALS

Singapore's recently submitted Intended Nationally Determined Contribution (INDC) to the Paris Agreement seeks to reduce greenhouse gas (GHG) emissions intensity 36 percent (from 2005 levels) by 2030 and stabilize emissions with the aim of peaking around 2030.[1] Given that more than 90 percent of Singapore's GHG emissions are from energy use, the INDC is a reasonable proxy for the country's clean energy goals. These goals aim to minimize the costs of reducing GHG emissions, in addition to a range of other policy objectives. However, as a densely populated city-state, constrained in both space and natural resources, Singapore faces unique challenges in achieving its clean energy goals.

On the supply side, natural gas is expected to remain the main fuel for power generation over the next few decades. Natural gas, combined with highly efficient combined cycle gas turbine power plant technology, accounts for 95 percent of the country's generated electricity.[2] Moreover, Singapore's high-capacity liquefied natural gas import terminal can tap into cost-competitive gas supplies from almost anywhere in the world, thus ensuring flexibility in and security of supply. In fact, the terminal is developing enough capacity for the country to become a regional hub for natural gas. Singapore's huge refining and petrochemical sector (largely based on petroleum) continues to pursue cleaner feedstock and energy options, including the use of energy-efficient cogeneration systems and environmentally benign bio-resources to produce clean renewable fuels in one of the largest bio-refineries in the world.

Other than its world-leading municipal waste-to-energy efforts (largely through mass-burn incinerators), Singapore's most promising domestic alternative energy source is the sun. Other low-carbon alternative energy options either cannot be practically harnessed (e.g., wind, marine renewables, biomass, hydropower) or are limited by safety concerns (in the case of nuclear energy) for a small city-state.

Singapore has introduced solar photovoltaic (PV) deployment initiatives over the last ten years, the public sector leading in the SolarNova program (using rooftops of public housing apartments, schools and government buildings, and reservoirs), which is projected to produce a 350 megawatt (MW) peak by 2020.[3] Plans are under way to use other potential spaces for PV deployment, including temporarily vacant land (with mobile solar PV systems), coastal waters (which could be extremely challenging considering Singapore's busy shipping and port activities), and building facades.[4] PV deployment is targeted to reach 1,000 MW by 2030. This capacity could satisfy around 15 percent of the country's current peak electricity demand and around 2.5 percent of total electricity demand.[5] Even these apparently modest contributions could result in Singapore's having one of the highest solar deployment intensities (MW deployed per square kilometer) in the world.

On the demand side, all sectors are aggressively pursuing energy conservation and efficiency. Already, Singapore's energy usage in the household, building, and transport sectors as well as some of its manufacturing subsectors are among the most efficient in the world, resulting in the country being ranked among the best performers globally in terms of carbon emissions intensity.[6] Policies and regulations such as the Energy Conservation Act, Green Building Master Plan and Green Mark scheme, Green Data Centre Standard, and Land Transport Master Plan strengthen the country's energy efficiency efforts. The planned introduction of an economy-wide carbon tax in 2019 could further motivate and strengthen efforts to achieve these goals.[7]

USING DIGITAL INNOVATIONS TO PROMOTE CLEAN ENERGY SYSTEMS

Even as a top-performing country in terms of energy and carbon emissions intensity, Singapore can do more. Digitalization in particular offers immense opportunities to transform energy systems; in Singapore, digitalization could maintain or improve desired levels of energy supply security, reliability, quality, flexibility, and resiliency while significantly mitigating environmental harm. It could reduce the cost of providing clean, sustainable energy and electricity to a society that already enjoys a highly reliable electricity system. Not all risks and trade-offs will disappear, but those that remain can be more intelligently and robustly assessed, managed, and solved thanks also to digitalization.

In line with this approach, Singapore has begun to invest substantially in research, development, and demonstration to incorporate digitalization in its energy system. Much of the current effort is funded under the ongoing Urban Solutions and Sustainability program of the Research Innovation and Enterprise 2020 plan, which builds on efforts of the last decade, such as the Energy Innovation Research Program and the Clean Energy Research Program.

Digitalization of the power grid, supported by the significant performance and cost improvements in technologies such as solar PV, batteries (for power grids as well as electric vehicles) and other energy storage technologies, and bidirectional solid-state power converters and transformers, will enable cost-competitive use and integration of variable renewable energy at much higher levels than Singapore's current target. As a result, it will be possible to make further progress toward decarbonization of Singapore's energy system without compromising other energy goals.

Digital technologies will enable more flexible and efficient power generation (from both conventional as well as new generation technologies), smart cross-sectoral grids (which provide not only electricity but also cooling and heat via multiple energy carriers), weather and solar radiation monitoring and forecasting, and high-resolution virtual city models that enable detailed spatial planning and design. Intelligent systems comprising smart sensors, meters and actuators, and robust and high-speed wired and wireless communications networks—which should support data- and information-sharing and be interoperable— together with the application of data science tools will enable the complex design, control, management, and optimization of the clean and smart energy system of the future.

On the demand side, multiple end-use application areas are already seeing their carbon footprints fall as a result of increased digitalization, including in homes and buildings and across the transportation and manufacturing sectors. Smart design tools for buildings and their energy systems (including air-conditioning and rooftop solar systems) are enabling even more improvements in energy-efficient building design. Smart control and management tools are enabling the optimal operation and management of building energy systems that take into account engineering knowledge of plant performance (e.g., efficiency of chillers at different loads), building space usage and occupancy, and external factors such as weather and outdoor air quality. In the future, smart transport systems could optimize energy efficiency through smart design, operation, and management of an integrated land transport system. Such a system would rely primarily on the mass rapid transit) and bus systems and derive ancillary support from mobility-sharing services (using electric and autonomous vehicles), and highway and road monitoring and management systems.

In many areas, energy supply and demand can no longer be treated as separate sectors to be independently optimized. For example, increased electrification of transport (e.g., as more electric vehicles are sold) and heating (as more customers use electric heat pumps) necessitate an integrated approach to designing and operating the energy system. The use of embedded cogeneration, integrated poly-generation, and multi-energy and low-carbon-energy district systems to enhance system efficiency, reduce overall costs, and minimize carbon emissions requires deep understanding both of energy supply and demand and of how the different technologies interact and integrate. Optimization at the systems level should take precedence over optimization of subsystems or individual technologies. Digital technologies will play a critical role in enabling systems optimization and overcoming inherent complexities and risks.

At the planning and design level, increasingly intelligent energy modeling and analysis techniques that incorporate techno-economic (bottom-up), macroeconomic (top-down), and integrated assessment approaches (e.g., considering wider land, water, resources, and climate issues) will provide important insights into extremely complex energy systems at national, sectoral (e.g., power systems, buildings, and transport sectors), city, and district levels. A strong understanding of energy demand services in terms of energy form and quality (e.g., electrical, heating, or cooling) and their temporal and spatial characteristics will be essential to enable the optimal matching and choice of candidate supply, transformation, distribution, and end-use conversion technologies to meet these demand services.

EMBEDDING ENERGY DIGITALIZATION INTO THE BROADER SMART NATION VISION

Complementing the efforts and initiatives specific to clean energy, the Singapore government, in partnership with the private sector and the civil society, has embarked on more general digitalization initiatives. The Smart Nation initiative is a useful case study of a comprehensive strategy to achieve rapid digitalization, in addition to making progress toward other energy and transport policy goals.[8]

Several overarching laws have been introduced to support the Smart Nation initiative. These include the Personal Data Protection Act of 2012, the Computer Misuse and Cybersecurity Act (amended in 2017), and new data-sharing and security provisions under the 2018 Public Sector (Governance) Bill. Policy agencies in Singapore recognize the challenge of enabling the use and disclosure of data to support technological progress and innovation while protecting personal data and privacy, and ensuring cybersecurity, especially in the government, banking, and energy sectors. Intensive efforts are under way to build capabilities in data analytics and cybersecurity that will support efforts in both the public and private sectors.

The government is implementing several infrastructure and innovation platforms and programs to support the goals of the Smart Nation initiative. These include the Next Generation Nationwide Broadband Network, which seeks to provide ultra-high-speed broadband access to all physical addresses; the Smart Nation Sensor Platform, a sensor and data collection initiative; Virtual Singapore, a 3D city modeling and collaborative data platform; and AI Singapore, an initiative to enhance the country's artificial intelligence capabilities.

To facilitate interoperability of digital devices and systems, standards for network system architecture, communications, and security protocols are being developed. Regulatory sandboxes, meant to be introduced for a limited duration and limited location or boundary across domains and sectors, are being tested outside existing regulations to facilitate learning, policy, and regulatory innovations.

LEADING REGIONALLY AND GLOBALLY ON DIGITALIZATION IN THE ENERGY SECTOR

Digitalization is playing an increasingly important role in Singapore's recent efforts to develop and use more clean energy. So far, the country has focused primarily on its own efforts, seeking cost-effective clean energy strategies and measures across the entire energy system. Developing a more sustainable and clean energy system to meet global challenges such as climate change and creating new business opportunities in clean energy technology will require Singapore to look beyond its borders—and not just for imports of fossil fuels.

Today, Singapore depends on energy imports because it cannot meet domestic energy demands with only domestic supplies. On the supply side, Singapore lacks substantial fossil fuel resources or the expanses of land needed for extensive solar PV deployment. On the demand side, in addition to the energy needs of its industrialized economy, Singapore also needs substantial energy to mitigate its water supply insecurity, for example by using energy-intensive desalination technology. As a result, Singapore is a substantial importer of fossil fuels from regional and global markets.

But through international cooperation, Singapore could tap into the significant clean energy resources of the Association of Southeast Asian Nations (ASEAN) region and beyond. Digitalization could help in planning, designing, analyzing, and subsequently managing and optimizing a complex regional energy system, such as the ASEAN Power

Grid, which is expected to enhance electricity trade across borders and provide benefits to meet the rising electricity demand and improve access to energy services in the region.[9]

Developments are encouraging. Singapore's government and public sector are willing to lead the various aforementioned initiatives and to help ASEAN build a network of smart cities and create new opportunities in the digital space. The private sector is also responding strongly—multinational corporations, large domestic enterprises, small and medium enterprises, and start-ups all recognizing the role of digitalization in enhancing business performance, energy efficiency, and energy management. Companies such as Siemens, Envision, Linde, Schneider Electric, and SPGroup are establishing centers of excellence and digitalization hubs that will serve not only Singapore but also the region and the world.

Singapore has thrived because of its openness to trade, investments, and talent. It has used strong political will and pragmatic analysis to forge bilateral and multilateral agreements with partners all over the world that benefit itself and its partners. Singapore can do the same in regional energy integration and collaboration beyond its borders, and digitalization could help achieve this.

Endnotes

INTRODUCTION

1. "The World's Most Valuable Resource Is No Longer Oil, but Data," *Economist*, May 6, 2017, http://economist.com/news/leaders/21721656-data-economy-demands-new-approach-antitrust-rules-worlds-most-valuable-resource.
2. International Energy Agency (IEA), *Digitalization and Energy* (Paris: IEA/OECD, 2017), http://iea.org/publications/freepublications/publication/DigitalizationandEnergy3.pdf.
3. Jason Bordoff, "How AI Will Increase the Supply of Oil and Gas—and Reduce Costs," *Wall Street Journal*, May 3, 2018, http://blogs.wsj.com/experts/2018/05/03/how-ai-will-increase-the-supply-of-oil-and-gas-and-reduce-costs.
4. Benjamin E. Gaddy et al. "Venture Capital and Cleantech: The Wrong Model for Energy Innovation," *Energy Policy* 102 (March 2017): 385–95.
5. Varun Sivaram, *The Dark Side of Solar: How the Rising Solar Industry Empowers Political Interests That Could Impede a Clean Energy Transition*, Brookings Institution, April 2018, http://brookings.edu/research/the-dark-side-of-solar.
6. IEA, *Digitalization and Energy*.
7. White House, *United States Mid-Century Strategy for Deep Decarbonization* (Washington, DC: Government Printing Office, November 2016), http://unfccc.int/files/focus/long-term_strategies/application/pdf/mid_century_strategy_report-final_red.pdf.
8. Daisuke Wakabayashi, "Self-Driving Uber Car Kills Pedestrian in Arizona, Where Robots Roam," *New York Times*, March 19, 2018, http://nytimes.com/2018/03/19/technology/uber-driverless-fatality.html.
9. Jacques Leslie, "Will Self-Driving Cars Usher in a Transportation Utopia or Dystopia?" *Yale Environment 360* (January 8, 2018), http://e360.yale.edu/features/will-self-driving-cars-usher-in-a-transportation-utopia-or-dystopia.
10. See previous competitions on Kaggle (http://kaggle.com/competitions).

TRENDS IN EARLY-STAGE FINANCING FOR CLEAN ENERGY INNOVATION

1. International Energy Agency, *Digitalization and Energy* (Paris: OEC/IEA, 2017), http://iea.org/publications/freepublications/publication/DigitalizationandEnergy3.pdf.
2. Updated version; original presented in Adam Bumpus and Stephen D. Comello, "Emerging Clean Energy Technology Investment Trends," *Nature Climate Change* 7, no. 6 (2017): 382–385.

3. Represents 72 percent of all venture capital transactions in this subsegment during this period.
4. Black & Veatch, 2017 *Strategic Directions: Electric Industry Report,* http://pages.bv.com/SDR-Electric-Industry-DL.html.
5. Bumpus and Comello, "Emerging Clean Energy Technology Investment Trends."
6. See IDEO CoLab, http://ideocolab.com; Elemental Excelerator, http://elementalexcelerator.com.

DIGITALIZATION: AN EQUAL OPPORTUNITY WAVE OF ENERGY INNOVATION

1. David G. Victor and Kassia Yanosek, "The Next Energy Revolution," *Foreign Affairs,* July/August 2017, http://foreignaffairs.com/articles/2017-06-13/next-energy-revolution.
2. Peter Thiel, *Zero to One: Notes on Startups, or How to Build the Future* (New York: Crown Business, 2017).
3. Devashree Saha and Mark Muro, "Cleantech Venture Capital: Continued Declines and Narrow Geography Limit Prospects," Brookings Institution, 2017, http://brookings.edu/research/cleantech-venture-capital-continued-declines-and-narrow-geography-limit-prospects.
4. Katie Fehrenbacher, "Venture Capital Funding for Cleantech Still Looks Pretty Grim," *Greentech Media,* May 16, 2017, http://greentechmedia.com/articles/read/venture-capital-funding-for-cleantech-still-looks-pretty-grim#gs.Od9w99w.
5. Jeffrey Ball et al., "The New Solar System," Steyer-Taylor Center for Energy Policy and Finance, March 2017, http://law.stanford.edu/publications/the-new-solar-system.
6. David M. Hart, "Across the 'Second Valley of Death': Designing Successful Energy Demonstration Projects," Information Technology and Innovation Foundation, July 26, 2017, http://itif.org/publications/2017/07/26/across-%22second-valley-death%22-designing-successful-energy-demonstration.
7. Adam Bumpus and Stephen D. Comello, "Emerging Clean Energy Technology Investment Trends," *Nature Climate Change* 7 (2017): 382–85.
8. Victor and Yanosek, "The Next Energy Revolution."
9. "The End of the Oil Age," *Economist,* October 23, 2003, http://economist.com/node/2155717.
10. Staffan Jacobsson and Volkmar Lauber, "The Politics and Policy of Energy System Transformation: Explaining the German Diffusion of Renewable Energy Technology," *Energy Policy* 34, no. 3 (2006): 256–76, http://doi.org/10.1016/j.enpol.2004.08.029.
11. Portions of this section are drawn from World Economic Forum, "Transformation of the Global Energy System: Riding the Innovation Tsunami," January 2018, http://www3.weforum.org/docs/White_Paper_Transformation_Global_Energy_System_report_2018.pdf.
12. See, for example, Meister Consultants Group, "Renewable Energy Revolution," March 16, 2015, http://mc-group.com/the-renewable-energy-revolution. The study argues that outlier groups were the most prescient predictors of the rise of solar energy.

A SURVEY OF DIGITAL INNOVATIONS FOR A DECENTRALIZED AND TRANSACTIVE ELECTRIC POWER SYSTEM

1. U.S.DepartmentofEnergy,"EconomicBenefitsofIncreasingGridResiliencetoWeather Outages," 2013, http://energy.gov/sites/prod/files/2013/08/f2/Grid%20Resiliency %20Report_FINAL.pdf.
2. World Economic Forum, "The Future of Electricity: New Technologies Transforming the Grid Edge," March 2017, http://www3.weforum.org/docs/WEF_Future_of _Electricity_2017.pdf.
3. GridWise Architecture Council, "GridWise Transactive Energy Framework," January 2015, http://gridwiseac.org/pdfs/te_framework_report_pnnl-22946.pdf.
4. David Livingston et al., "Blockchain and Electric Power Systems," Council on Foreign Relations (forthcoming).

HOW DISTRIBUTION ENERGY MARKETS COULD ENABLE A LEAN AND RELIABLE POWER SYSTEM

1. Herman K. Trabish, "Why NARUC Wants State Regulators to Incentivize Utility Cloud Computing," *Utility Dive*, December 7, 2016, http://utilitydive.com/news /why-naruc-wants-state-regulators-to-incentivize-utility-cloud-computing/431603.
2. Ben Foster, "Assessment of Demand Response and Advanced Metering," Federal Energy Regulatory Commission, December 2017, http://ferc.gov/legal/staff-reports /2017/DR-AM-Report2017.pdf.
3. Jeff St. John, "PJM's Latest Capacity Auction: A Tough Market for Nuclear and Demand Response," Greentech Media, May 24, 2017, http://greentechmedia.com /articles/read/pjms-capacity-auction-a-poor-showing-for-nuclear-and-demand -response#gs.V2bt64I; "Optimizing the Demand Response Enrollment Process," EnergyHub, http://energyhub.com/optimizing-demand-response-enrollment.
4. William W. Hogan, "Markets in Real Electric Networks Require Reactive Prices," *Energy Journal* 14, no. 13 (1993): 171–200, http://jstor.org/stable/41322516.
5. Richard Tabors et al., "White Paper on Developing Competitive Electricity Markets and Pricing Structures," Tabors Caramanis Rudkevich, April 2016, http://sites.hks.harvard.edu/hepg/Papers/2016/TCR.%20White%20Paper%20on %20Developing%20Competitive%20Electricity%20Markets%20and%20Pricing %20Structures.pdf.
6. Paul M. Sotkiewicz and Jesus M. Vignolo, "Nodal Pricing for Distribution Networks: Efficient Pricing for Efficiency Enhancing DG," *IEEE Transactions on Power Systems* 21, no. 21 (2006): 1013–1014, http://iie.fing.edu.uy/publicaciones/2006/SVo6b/VSo6.pdf.
7. Maher Alharby and Aad van Moorsel, "Blockchain-Based Smart Contracts: A Systematic Mapping Study," *Fourth International Conference on Computer Science and Information Technology* (2017), http://arxiv.org/ftp/arxiv/papers/1710/1710.06372.pdf.
8. Mike Orcutt, "How Blockchain Could Give Us a Smarter Energy Grid," *MIT Technology Review*, October 16, 2017, http://technologyreview.com/s/609077/how-blockchain -could-give-us-a-smarter-energy-grid.

9. Emma Nicholson, "Operator-Initiated Commitments in RTO and ISO Markets," Federal Electricity Regulatory Commission, December 2014, http://ferc.gov/legal/staff-reports/2014/AD14-14-operator-actions.pdf.

10. Paul R. Gribik, William W. Hogan, and Susan L. Pope, "Market-Clearing Electricity Prices and Energy Uplift," December 31, 2007, http://sites.hks.harvard.edu/fs/whogan/Gribik_Hogan_Pope_Price_Uplift_123107.pdf.

11. Elli Ntakou and Michael Caramanis, "Distribution Network Electricity Market Clearing: Parallelized PMP Algorithms with Minimal Coordination," IEEE 53rd Annual Conference on Decision and Control (December 15–17, 2014), http://ieeexplore.ieee.org/abstract/document/7039642.

THE IMPLICATIONS OF VEHICLE ELECTRIFICATION AND AUTONOMY FOR GLOBAL DECARBONIZATION

1. Alexis Madrigal, "All the Promises Automakers Have Made about the Future of Cars," *Atlantic*, July 7, 2017, http://theatlantic.com/technology/archive/2017/07/all-the-promises-automakers-have-made-about-the-future-of-cars/532806; Brett Williams, "If It Seems Like Every Carmaker Is Going Electric, That's Because They Are," Mashable, October 3, 2017, http://mashable.com/2017/10/03/electric-car-development-plans-ford-gm.

2. Jeffrey Mervis, "Not So Fast," Science 358, no. 6369 (December 15, 2017): 1370–74, http://science.sciencemag.org/content/358/6369/1370.

3. U.S. Energy Information Agency, "Energy Use for Transportation," last updated May 17, 2017, http://eia.gov/energyexplained/index.cfm?page=us_energy_transportation#tab1.

4. U.S. Energy Information Agency, "Power Sector Carbon Dioxide Emissions Fall Below Transportation Sector Emissions," January 19, 2017, http://eia.gov/todayinenergy/detail.php?id=29612.

5. The International Energy Agency estimates that global energy demand will increase 30 percent by 2040 in its baseline New Policies scenario. Continued population and economic growth beyond 2040 will bring total growth to 50 percent. See International Energy Agency, *World Energy Outlook* 2017 (Paris: OECD/IEA, November 14, 2017), http://iea.org/weo2017.

6. A vehicle in the U.S. fleet averages about twenty-five miles per gallon. At an emissions factor of roughly twenty pounds of carbon dioxide (CO_2) per gallon, an ICE vehicle emits around eighty pounds of CO_2 per hundred miles traveled. Electric vehicles, on the other hand, use around 34 kWh of electricity per hundred miles traveled. For electric vehicles to produce fewer emissions than their ICE counterparts, the electricity system needs to generate less than 2.35 pounds of CO_2 per kWh delivered. For reference, the average coal-fired power plant in the United States generates 2 to 2.2 pounds of CO_2 per kWh. Without transmission losses, even a 100 percent coal-fired grid would reduce emissions. The United States is well below that emissions figure: roughly 1.13 pounds of CO_2 are generated per kWh (see "1 Kilowatt-Hour," BlueSkyModel, http://blueskymodel.org/kilowatt-hour). China's power sector emits 1.56 pounds per kWh. For a selection of different country emissions factors, see Vivien Foster and Daron Bedrosyan, "Understanding CO_2 Emissions from the Global Energy Sector," World Bank, 2014, http://documents.worldbank.org/curated/en/873091468155720710/pdf/851260BRI0Live00Box382147B00PUBLIC0.pdf.

7. "What Cheap, Clean Energy Means for Global Utilities," Morgan Stanley Research, July 5, 2017.
8. Eric Wood et al., "National Plug-In Electric Vehicle Infrastructure Analysis," National Renewable Energy Laboratory, September 2017, http://nrel.gov/docs/fy17osti/69031.pdf.
9. Jürgen Weiss et al., "The Electrification Accelerator: Understanding the Implications of Autonomous Vehicles for Electric Utilities," *Electricity Journal* 30, no. 10 (December 2017): 50–57, http://doi.org/10.1016/j.tej.2017.11.009.
10. In the developed economies of the world, this assumes that energy use for bitcoin mining, which has grown rapidly in the last years, levels off or declines.
11. Peter Fox-Penner, Will Gorman, and Jennifer Hatch, "Long-Term Transportation Electricity Use: Estimates and Policy Ob-servations" (working paper, Institute for Sustainable Energy, Boston University, Boston, MA, 2017) .
12. Jules Polonetsky and Henry Claypool, "Self Driving Cars: Transforming Mobility for the Elderly and People with Disabilities," *Huffington Post*, last updated October 29, 2017, http://huffingtonpost.com/julespolonetsky/selfdriving-cars-transfor_b_12545726.html.
13. James Arbib and Tony Seba, "Rethinking Transportation 2020–2030," RethinkX, 2017, https://static1.squarespace.com/static/585c3439be65942f022bbf9b/t/591a2e4be6f2e1c13df930c5/1494888038959/RethinkX+Report_051517.pdf.
14. See, for example, Maria Kamargianni et al., "A Critical Review of New Mobility Services for Urban Transport," *Transportation Research Procedia* 14 (2016): 3294–303, http://doi.org/10.1016/j.trpro.2016.05.277; and "New York City Taxi and Limousine Commission 2013 Annual Report," http://nyc.gov/html/tlc/downloads/pdf/annual_report_2013.pdf.
15. New York City Taxi and Limousine Commission, "2013 Annual Report," http://nyc.gov/html/tlc/downloads/pdf/annual_report_2013.pdf.
16. See, for example, Daniel J. Fagnant and Kara Kockelman, "Preparing a Nation for Autonomous Vehicles: Opportunities, Barriers and Policy Recommendations," *Transportation Research* A 77 (2015): 167–81.
17. Tim Higgins, "The End of Car Ownership," *Wall Street Journal*, June 20, 2017, http://wsj.com/articles/the-end-of-car-ownership-1498011001.
18. See, generally, the work of Professor Christos Cassandras, Boston University, such as Yue Z. Zhang, Andreas A. Malikopoulos, and Cassandras, "Optimal Control of Connected and Automated Vehicles at Urban Traffic Intersections" (paper, American Control Conference, Boston, MA, July 6–8, 2016), http://ieeexplore.ieee.org/document/7526648.
19. Intel, "Autonomous Cars: The Road Ahead," August 16, 2016, http://iq.intel.com/autonomous-cars-road-ahead.
20. Ohio Department of Transportation, "Smart Mobility Corridor to Become Ohio's First 'Smart Road,'" November 30, 2016, http://www.dot.state.oh.us/news/Pages/SmartMobilityCorridor.aspx.
21. American Society of Civil Engineers, "2017 Infrastructure Report Card: Roads," 2017, http://infrastructurereportcard.org/wp-content/uploads/2017/01/Roads-Final.pdf.
22. Victor M. Mendez, Carlos A. Monje, and Vinn White, "Beyond Traffic: Trends and Choices 2045; A National Dialogue About Future Transportation Opportunities and Challenges," in *Disrupting Mobility: Impacts of Sharing Economy and Innovative Transportation on Cities*, ed. Gereon Meyer and Susan Shaheen (New York: Springer, 2016).
23. Fox-Penner, Gorman, and Hatch, "Long-Term Transportation Electricity Uses."
24. Fox-Penner, Gorman, and Hatch, "Long-Term Transportation Electricity Uses."

AUTONOMOUS VEHICLES AND CITIES: EXPECTATIONS, UNCERTAINTIES, AND POLICY CHOICES

1. Kasey Panetta, "Top Trends in the Gartner Hype Cycle for Emerging Technologies, 2017," Gartner, August 15, 2017, http://gartner.com/smarterwithgartner/top-trends -in-the-gartner-hype-cycle-for-emerging-technologies-2017.

2. Robin Chase, "Self-Driving Cars Will Improve Our Cities. If They Don't Ruin Them," *Wired*, August 10, 2016, http://wired.com/2016/08/self-driving-cars-will-improve-our -cities-if-they-dont-ruin-them.

3. Alexis C. Madrigal, "The Most Important Self-Driving Car Announcement Yet," *Atlantic* (online), March 28, 2018, http://theatlantic.com/technology/archive/2018/03 /the-most-important-self-driving-car-announcement-yet/556712. Autonomous vehicles will transform urban life by 2020, if Waymo's time line is correct; Marshall, "Waymo Launches Its Self-Driving Armada," *Wired*, January 30, 2018, http://wired.com/story /waymo-launches-self-driving-minivans-fiat-chrysler; James Titcomb, "Uber Boss Says Driverless Cars Are at Least a Decade Away," *Telegraph*, January 22, 2018, http://telegraph.co.uk/technology/2018/01/22/uber-boss-says-driverless-cars-least -decade-away; Daisuke Wakabayashi, "Self-Driving Uber Car Kills Pedestrian in Arizona, Where Robots Roam," *New York Times*, March 19, 2018, http://nytimes .com/2018/03/19/technology/uber-driverless-fatality.html.

4. Neal E. Boudette, "G.M. Says Its Driverless Car Could Be in Fleets by Next Year," *New York Times*, January 12, 2018, http://nytimes.com/2018/01/12/business/gm-driverless -car.html; Timothy J. Seppala, "Waymo's Driverless Taxi Service Will Open to the Public Soon," Engadget, November 7, 2017, http://engadget.com/2017/11/07 /waymo-autonomous-taxi-phoenix.

5. Johana Bhuiyan, "Why GM Is Vertically Integrating as It Moves Deeper into Making Self-Driving Cars," Recode, October 9, 2017, http://recode.net/2017/10/9/16446768 /general-motors-acquisition-lidar-sensor-startup-strobe-self-driving.

6. See National Highway Transportation Safety Administration, "National Motor Vehicle Crash Causation Survey: Report to Congress," DOT GS 811059, July 2008, http://crashstats.nhtsa.dot.gov/Api/Public/ViewPublication/811059. The frequently cited 95 percent statistic is based on the percent of all crashes attributed to driver error in the 2008 study by the National Highway Transportation Safety Administration. That report, however, includes a category of bad judgment, such as misjudging of other drivers' intentions and inaccurately interpreting road conditions that AV technology will not necessarily fix. Excluding this and including only the obvious driver failures such as distracted driving and falling asleep at the wheel, it is likely that AVs could reduce crashes by 60 percent.

7. Bruce Schaller, "Empty Seats, Full Streets: Fixing Manhattan's Traffic Problem," December 21, 2017, http://schallerconsult.com/rideservices/emptyseats.pdf.

8. Andrew J. Hawkins, "Not All of Our Self-Driving Cars Will Be Electrically Powered— Here's Why," *Verge*, December 12, 2017, http://theverge.com/2017/12/12/16748024 /self-driving-electric-hybrid-ev-av-gm-ford.

9. International Transport Forum, "Urban Mobility System Upgrade: How Shared Self-Driving Cars Could Change City Traffic," March 31, 2015, http://itf-oecd.org/sites /default/files/docs/15cpb_self-drivingcars.pdf.

10. Josh Stephens, "How One City Will Change Its Entire Bus System Overnight," Next City, July 24, 2015, http://nextcity.org/daily/entry/houston-change-entire-bus -system-overnight.

11. See the Regional Plan Association's proposal to do this for New York City's subways. Regional Plan Association, *The Fourth Regional Plan: Making the Region Work for All of Us*, November 2017, http://library.rpa.org/pdf/RPA-The-Fourth-Regional-Plan.pdf.

12. Jim Dwyer, "Volkswagen's Diesel Fraud Makes Critic of Secret Code a Prophet," *New York Times*, September 22, 2015, http://nytimes.com/2015/09/23/nyregion /volkswagens-diesel-fraud-makes-critic-of-secret-code-a-prophet.html.

13. James Temple, "Alphabet's Moonshot Chief: Regulating Driverless Cars Demands Testing the 'Smarts' of the Systems," *MIT Technology Review*, March 27, 2018, http:// technologyreview.com/s/610680/alphabets-moonshot-chief-regulating-driverless -cars-demands-testing-the-smarts-of-the.

14. David Welch and Gabrielle Coppola, "Self-Driving Taxis Could Have a Vomit Problem," Bloomberg, July 13, 2017, http://bloomberg.com/news/articles/2017-07-13 /robotaxis-vomit-problem-there-s-no-one-to-clean-up-the-mess.

15. Chase, "Self-Driving Cars Will Improve Our Cities."

16. Gustavo Vicentini, "Location Strategy of Chain Retailers: The Case of Supermarkets and Drug Stores in an Urban Market," October 2012, http://goo.gl/7EfJQU.

17. See International Transport Forum, *Shared Mobility Solutions for Auckland* (Paris: OECD/ITF, 2017), 64, http://itf-oecd.org/sites/default/files/docs/shared-mobility-simulations-auckland.pdf. This study on applying shared-ride services to low-density, auto-dependent Auckland, New Zealand, showed that in every scenario, pushing any portion of the 82 percent of the trips currently made by private automobiles results in an increase in the overall travel time for those users.

HOW DATA SCIENCE CAN ENABLE
THE EVOLUTION OF ENERGY SYSTEMS

1. Adam Cooper, "Electric Company Smart Meter Deployments: Foundation for a Smart Grid," Institute for Electric Innovation, October 2016, http://edisonfoundation .net/iei/publications/Documents/Final%20Electric%20Company%20Smart %20Meter%20Deployments-%20Foundation%20for%20A%20Smart%20Energy %20Grid.pdf.

2. Alex Krizhevsky, Ilya Sutskever, and Geoffrey E. Hinton, "ImageNet Classification With Deep Convolutional Neural Networks," *Advances in Neural Information Processing Systems* 25 (2012): 1097–105, http://papers.nips.cc/paper/4824-imagenet -classification-with-deep-convolutional-neural-networks.pdf.

3. U.S. Energy Information Administration, *Electric Power Monthly with Data for August 2017* (Washington, DC: U.S. Department of Energy, October 2017), http://eia.gov /electricity/monthly/archive/october2017.pdf.

4. K. Carrie Armel et al., "Is Disaggregation the Holy Grail of Energy Efficiency? The Case of Electricity," Energy Policy 52 (January 2013): 213–34, http://web.stanford .edu/~adalbert/papers/disaggregation_paper.pdf.

5. Joseph Camilo et al., "Application of a Semantic Segmentation Convolutional Neural Network for Accurate Automatic Detection and Mapping of Solar Photovoltaic Arrays in Aerial Imagery"(paper presentation, 46th Annual IEEE Applied Imagery Pattern Recognition Workshop, Washington, DC, October 10–12, 2017).

APPLYING DATA SCIENCE
TO PROMOTE RENEWABLE ENERGY

1. Lazard, "Levelized Cost of Energy 2017," November 2, 2017, http://lazard.com/perspective
   /levelized-cost-of-energy-2017.
2. Consolidated Appropriations Act, 2016, Public Law No. 114-113, 129 Stat. 2242 (2015).
3. U.S. Energy Information Administration, "Almost All Power Plants that Retired in the
   Past Decade Were Powered by Fossil Fuel," *Today in Energy*, January 9, 2018, http://eia
   .gov/todayinenergy/detail.php?id=34452.
4. Robert Rice and Richard Bontatibus, "Predictive Maintenance Embraces Analytics,"
   InTech Magazine, January/February 2013, http://isa.org/standards-publications/isa
   -publications/intech-magazine/2013/feb/automation-it-predictive-maintenance
   -embraces-analytics.
5. Uptake Technologies, "Untapped Energy: How the U.S. Wind Industry Can Produce
   More Megawatts without New Turbines," 2017, http://uptake.com/untapped
   -energy-report.
6. Richard J. Campbell, "Increasing the Efficiency of Existing Coal Fired Power Plants,"
   CRS Report no. R.43343, December 20, 2013, http://fas.org/sgp/crs/misc/R43343.pdf.
7. See *Permian Basin Area Rate Cases*, 390 U.S. 747 (1968); *Bluefield Co. v. Pub. Serv.
   Comm.*, 262 U.S. 679 (1923); *Power Comm'n v. Hope Gas Co.*, 320 U.S. 591 (1944).
8. Karl McDermott, "Cost of Service Regulation In the Investor-Owned Electric Utility
   Industry: A History of Adaptation," Edison Electric Institute, June 2012, http://eei.org
   /issuesandpolicy/stateregulation/Documents/COSR_history_final.pdf.
9. National Association of Regulatory Utility Commissioners, "Resolution Encouraging
   State Utility Commissions to Consider Improving the Regulatory Treatment of
   Cloud Computing Arrangements," November 16, 2016, http://pubs.naruc.org/pub
   .cfm?id=2E54C6FF-FEE9-5368-21AB-638C00554476.
10. Next Generation GRID Act, S. 2232, 115 Cong. (2017).
11. Illinois Commerce Commission, "Notice of Inquiry Regarding the Regulatory
    Treatment of Cloud-Based Solutions," April 7, 2017, http://icc.illinois.gov/downloads
    /public/NOI%20Regulatory%20Treatment%20of%20CloudBased%20Solutions%20
    -%20Report.docx.
12. State of New York Public Service Commission, "Order Adopting a Ratemaking and Utility
    Revenue Model Policy Framework," May 19, 2016, http://documents.dps.ny.gov
    /public/Common/ViewDoc.aspx?DocRefId=%7bD6EC8F0B-6141-4A82-A857
    -B79CF0A71BF0%7d.
13. Digital Millennium Copyright Act, 17 U.S.C. § 105 (1998).
14. Circumvention of Copyright Protection Systems, 17 U.S.C. § 1201(a)(1) (2015) .
15. Institute of Nuclear Power Operations , "About Us," accessed February 9, 2018. http://
    inpo.info/AboutUs.htm.

MANAGING THE ECONOMIC
AND PRIVACY RISKS ARISING
FROM DIGITAL INNOVATIONS IN ENERGY

1. International Energy Agency, *Digitalization and Energy 2017* (Paris: OECD/IEA, 2017), http://iea.org/publications/freepublications/publication/DigitalizationandEnergy3.pdf.

2. Joint Utilities, *Supplemental Distributed System Implementation Plan*, Case 14-M-0101 (New York: Joint Utilities, November 2016), http://jointutilitiesofny.org/wp-content/uploads/2016/10/3A80BFC9-CBD4-4DFD-AE62-831271013816.pdf.

3. Eurelectric, "The Power Sector Goes Digital: Next Generation Data Management for Energy Consumers" (Brussels: Union of the Electricity Industry, 2016), http://www3.eurelectric.org/media/278067/joint_retail_dso_data_report_final_11may_as-2016-030-0258-01-e.pdf.

4. Andrew Bartholomew, "The Smart Grid in Massachusetts: A Proposal for a Consumer Data Privacy Policy," *Boston College Environmental Affairs Law Review* 43, no. 1 (2016): 79–110, http://lawdigitalcommons.bc.edu/ealr/vol43/iss1/4.

5. Federal Republic of Germany, Messstellenbetriebsgesetz vom 29 August 2016 [Metering Point Operating Act of 29 August 2016], http://gesetze-im-internet.de/messbg/BJNR203410016.html; for France, see Commission Nationale de l'Informatique et des Libertés [National Commission for Information Technology and Rights], http://www.cnil.fr/en/home.

6. U.S. Department of Energy, "Voluntary Code of Conduct (VCC): Final Concepts and Principles," January 12, 2015, http://energy.gov/sites/prod/files/2015/01/f19/VCC%20Concepts%20and%20Principles%202015_01_08%20FINAL.pdf.

7. UK Information Commissioner's Office, "The Information Commissioner's Office Response to the Department of Business, Energy & Industrial Strategy and Ofgem's Call for Evidence on 'a Smart, Flexible Energy System,'" January 2017, http://ico.org.uk/media/about-the-ico/consultation-responses/2017/1625738/ico-response-beis-smart-energy-20170112.pdf.

8. Republic of France, Article 19 of Law No. 2016-1321 of October 7, 2016, for a Digital Republic (1), http://legifrance.gouv.fr/affichTexte.do?cidTexte=JORFTEXT000033202746&dateTexte=20180323.

9. California Civil Code, Division 3: Part 4: Obligations arising from particular transactions. See also draft bill SB-356 on energy data transparency.

10. Smart Grid Act 2011, 지능형전력망의 구축 및 이용촉진에 관한 법률/知能型電力網法.

11. International Energy Agency, *World Energy Investment 2017* (Paris: OECD/IEA, 2017), http://iea.org/publications/wei2017.

12. International Transport Forum, "Managing the Transition to Driverless Road Freight Transport," 2017, http://itf-oecd.org/sites/default/files/docs/managing-transition-driverless-road-freight-transport.pdf.

13. European Commission, "Digital Single Market: Bringing Down Barriers to Unlock Online Opportunities," http://ec.europa.eu/commission/priorities/digital-single-market_en#documents.

14. International Energy Agency, "Technology Collaboration Programmes: Highlights and Outcomes," 2018, http://iea.org/tcp.

HOW STATE-LEVEL REGULATORY REFORM CAN ENABLE THE DIGITAL GRID OF THE FUTURE

1. James D. Dana Jr. and Eugene Orlov, "Internet Penetration and Capacity Utilization in the US Airline Industry," *American Economic Journal: Microeconomics* 6, no. 4 (November 2014): 106–37.

LESSONS FROM SINGAPORE'S APPROACH TO DEVELOPING CLEAN AND DIGITAL ENERGY SYSTEMS

1. UN Framework Convention on Climate Change Secretariat, "Singapore's Intended Nationally Determined Contribution (INDC) and Accompanying Information," July 3, 2015, http://www4.unfccc.int/Submissions/INDC/Published%20Documents/Singapore/1/Singapore%20INDC.pdf.
2. Energy Market Authority, "Singapore Energy Statistics," 2017, http://ema.gov.sg/cmsmedia/Publications_and_Statistics/Publications/SES17/Publication_Singapore_Energy_Statistics_2017.pdf.
3. National Climate Change Secretariat (NCCS), "Take Action Today: For a Carbon-Efficient Singapore" (Singapore: NCCS, 2016), http://nccs.gov.sg/docs/default-source/publications/take-action-today-for-a-carbon-efficient-singapore.pdf.
4. Ministry of Communications and Information, "Singapore Energy Lecture cum Dialogue by Mr Teo Chee Hean, Deputy Prime Minister," Singapore, 2017, http://gov.sg/~/sgpcmedia/media_releases/mcicad/speech/S-20171023-1/attachment/Media%20copy%20-%20231017_Speech%20by%20DPM%20Teo%20at%20SIEW%202017.pdf.
5. Prime Minister's Office, "DPM Teo Chee Hean at the opening ceremony of the Singapore Sustainability Academy," Press Release, June 5, 2017, http://pmo.gov.sg/newsroom/dpm-teo-chee-hean-opening-ceremony-singapore-sustainability-academy.
6. NCCS, Strategy Group Prime Minister's Office, "Singapore's Emissions Profile," last updated January 19, 2018, http://nccs.gov.sg/climate-change-and-singapore/national-circumstances/singapore's-emissions-profile.
7. NCCS, Strategy Group Prime Minister's Office, "Carbon Tax," last updated February 19, 2018, https://www.nccs.gov.sg/climate-change-and-singapore/reducing-emissions/carbon-tax.
8. International Energy Agency, *Digitalization and Energy* (Paris: OECD/IEA, 2017), http://iea.org/publications/freepublications/publication/DigitalizationandEnergy3.pdf.
9. ASEAN Centre for Energy, "ASEAN Power Grid," 2017, http://aseanenergy.org/programme-area/apg.

Acknowledgments

I am indebted to James M. Lindsay, senior vice president and director of Studies at the Council on Foreign Relations (CFR), for his support for this project over the course of more than a year. When I first presented the idea to convene experts to discuss how digital technologies such as machine learning were transforming the energy landscape, he urged me to compile and publish the insights; *Digital Decarbonization* is the result. CFR President Richard N. Haass was supportive and encouraged me to address the implications of transformative digital technologies.

This workshop was the sixth in a series of energy-focused gatherings supported by the Alfred P. Sloan Foundation. I am extremely grateful to Evan Michelson, who served as a sounding board for the workshop's themes. I admire Evan's commitment to cultivating a community of scholars studying energy innovation, and his support has made possible a rich set of insights and interactions.

I am grateful to the diverse collection of experts who joined the workshop and shared their insights. I am especially thankful to the contributors to this volume, who presented at the workshop and subsequently worked with me on editing their essays. I am particularly appreciative that they took the time to go through as many as nine rounds of editing to incorporate responses to other essays and ensure that the various pieces of the manuscript assembled into a coherent whole.

Finally, I want to thank the many team members at CFR who made *Digital Decarbonization* possible. Patricia Dorff and Sumit Poudyal combed through the manuscript to make it more accessible, consistent, and precise. Amanda Shendruk created engaging graphics. And a special note of thanks is due to my research associate, Madison Freeman. Madison spearheaded every detail of the workshop, from inviting participants to arranging logistics, and her research was invaluable in shaping the workshop agenda and writing up the resulting insights. She also helped edit, format, and cite all of the essays in this volume, and managed our two excellent interns, David Yellen and Max Fiege, to pull off the workshop and this publication.

Varun Sivaram
June 2018

About the Authors

Rohit T. Aggarwala is the head of urban systems at Sidewalk Labs, chair of the Regional Plan Association's fourth regional plan for the New York metropolitan area, and an adjunct professor in Columbia University's School of International and Public Affairs. He previously served in multiple roles, developing the sustainability and environmental practices of Bloomberg Associates and Bloomberg Philanthropies, as well as serving as the director of long-term planning and sustainability for New York during Michael Bloomberg's mayoralty. He also acted as special advisor to the chair in the C40 Cities Climate Leadership Group. He was previously a management consultant at McKinsey & Company, where his practice focused on transportation and telecommunications. Aggarwala earned a BA in history from Columbia University, an MA in Canadian history from Queen's University, an MBA in finance from Columbia Business School, and a PhD in U.S. history from Columbia University.

Kyle Bradbury is the managing director of the Energy Data Analytics Lab at the Duke University Energy Initiative, where he leads applied research projects at the intersection of machine learning techniques and energy problems. His research includes developing techniques for automatically mapping global energy infrastructure and access from satellite imagery; transforming smart electric utility meter data into energy efficiency insights; and exploring the reliability and cost trade-offs of energy storage systems for integrating wind and solar power into the grid. Prior to working at Duke, Bradbury worked for ISO New England, MIT Lincoln Laboratory, and Dominion Energy. Bradbury received a BS in electrical engineering from Tufts University and an MS in electrical and computer engineering and a PhD in energy systems modeling from Duke.

Stephen D. Comello is the director of the Sustainable Energy Initiative at Stanford Graduate School of Business, a founding researcher at the Bits & Watts Initiative within Stanford University's Precourt Institute for Energy, and a fellow at the Steyer-Taylor Center for Energy Policy and Finance. His work examines the organization of innovation and how technology and policy coevolve to influence the economic attractiveness of advanced energy solutions. He advises academic, industry, government and nongovernmental organizations on strategies for clean technology deployment. His current portfolio explores policy and business model innovation within the electricity sector, with one set of projects focusing on how digital platform technology diffusion intersects with structural changes in the industry. Comello received his PhD in civil and environmental engineering, with a PhD minor in management science and engineering, from Stanford.

Peter Fox-Penner is a professor of practice in the Questrom School of Management and the director of Boston University's Institute for Sustainable Energy. In addition, he is chief strategy officer of Energy Impact Partners and academic advisor to the Brattle Group and serves on the advisory board of EOS Energy Storage. He researches and writes on electric power strategy, regulation, and governance; energy and climate policy; and the relationships between public and private economic activity. The author of *Smart Power: Climate Change, the Smart Grid, and the Future of Electric Utilities*, as well as other books in this area, Fox-Penner also teaches courses on sustainable energy and electric power in the Questrom School. Fox-Penner received his BS and MS from the University of Illinois and his PhD from the University of Chicago.

Sunil Garg is the global managing director for energy solutions at Uptake Technologies, a machine learning and artificial intelligence software company committed to improving the productivity, reliability, safety and security of industrial assets. Prior to joining Uptake, Garg was the chief information and innovation officer at Exelon Corp and served as president of Exelon Power. In the public sector, he served Mayor Richard Daley of Chicago and was appointed a White House fellow by President Bill Clinton. Garg received his BA from the University of Chicago, his MPP from the Harvard Kennedy School, and his MBA from the University of Chicago Booth School of Business.

Ben Hertz-Shargel is vice president of analytics at EnergyHub, responsible for the company's Mercury DERMS platform and energy market operations. As the company's lead data scientist, his primary focus is modeling and optimal control of distributed energy resources for distribution, market, and customer grid services. Hertz-Shargel was previously vice president of technology for ThinkEco. Prior to that, he held applied research positions in the quantitative strategies group at Credit Suisse, the New England Complex Systems Institute, the physics department at the Massachusetts Institute of Technology (MIT), and Icosystem Corporation. Hertz-Shargel holds a BA in computer science from Northwestern University, an MS in mathematics from the New York University's Courant Institute, and a PhD in mathematics from the University of California, Los Angeles.

Erfan Ibrahim is the founder and CEO of the Bit Bazaar LLC. Previously, he worked for the National Renewable Energy Lab, where he led the Cyber-Physical Systems Security and Resilience Center for three years and pioneered the concept of a nine-layer cybersecurity architecture to protect power systems from insider and external cyber threats. Before that, Ibrahim worked at Lawrence Livermore National Lab; University of California, Los Angeles; Pacific Bell, Jyra Research, and Electric Power Research Institute. Ibrahim received a BS in physics from Syracuse University, an MS in mechanical engineering from University of Texas at Austin, and a PhD in nuclear engineering from University of California, Berkeley.

Richard Kauffman serves as New York State's first energy czar (officially the chairman of energy and finance for New York), appointed by Governor Andrew M. Cuomo. In his position, he oversees the state's energy agencies, including the Department of Public Service, the New York Power Authority, the Long Island Power Authority, and the New York State Energy Research and Development Authority, where he also serves as chair. Kauffman leads New York State's Reforming the Energy Vision (REV) policy. REV is a market-based approach to build an electric system that is cleaner and more affordable by changing utility financial incentives and reforming the state's clean energy support mechanisms. Kauffman previously served as senior advisor to U.S. Secretary of Energy Steven Chu, was president and CEO of Good Energies, was a leading investor in clean energy, and held senior roles in finance as a partner at Goldman Sachs and as vice chairman of institutional securities at Morgan Stanley. Kauffman received a bachelor's degree from Stanford University and master's degrees from Yale University.

Hiang Kwee Ho is an adjunct associate professor at Nanyang Technological University's School of Mechanical and Aerospace Engineering, in Singapore. He is also lead technologist in the strategy group of the National Climate Change Secretariat in the Singaporean prime minister's office. Ho's technical interests and work in government and academia are in sustainable energy and climate change mitigation technologies and systems, and include energy efficiency technologies, conventional power generation technologies, fuel cells, carbon capture, storage and utilization, integrated energy systems, and the application of digitalization in the modeling, analysis, design, operation, management, and optimization of energy systems. Ho received his BSc from the University of Newcastle upon Tyne and his MS from the Massachusetts Institute of Technology.

John O'Leary is senior policy advisor for energy and finance for New York State. He joined the Andrew M. Cuomo administration in February 2014 and is involved in developing and executing all aspects of the Reforming the Energy Vision agenda. Following Cuomo's founding of the U.S. Climate Alliance in June 2017, O'Leary helps lead New York State's efforts to broaden its energy policy approaches to other states, including the national expansion of NY Green Bank. Previously, he worked on renewable energy policy and research at the American Council on Renewable Energy in Washington, DC. O'Leary holds a BA from George Washington University.

Jesse Scott is the deputy secretary-general of the business association Eurogas, in Brussels. She was formerly an analyst for electricity markets and digitalization for the International Energy Agency (IEA), where she served as the lead author of the IEA's 2017 report *Energy and Digitalization*. Before joining the IEA, she led the policies and decarbonization team at Eurelectric, was the program director for energy and climate for demosEuropa, and was head of the EU office of the environmental group E3G. Scott holds a BA in history and an MPhil in European studies from Cambridge University.

Lidija Sekaric is the director of strategy and marketing of distributed energy systems at Siemens. Prior to joining Siemens, she served as a director, deputy director, and group manager in the Department of Energy's SunShot Initiative, managing solar research and development funding and driving the department's solar energy strategy. She concurrently served as the senior technical advisor to the undersecretary for energy, directing strategic projects to implement various energy technologies, including renewables and nuclear power. Sekaric began her career as a research scientist in IBM's nanostuctures and exploratory devices group, where she conducted research into nanophotonics and semiconductor development. She received her BA in physics from Bryn Mawr College and her PhD in applied physics from Cornell University.

Varun Sivaram is the Philip D. Reed fellow for science and technology at the Council on Foreign Relations. He is also an adjunct professor at Georgetown University's Edmund A. Walsh School of Foreign Service, a nonresident fellow at Columbia University's Center for Global Energy Policy, and a member of the advisory boards for Stanford University's Woods Institute for the Environment and Precourt Institute for Energy. He is the author of the book *Taming the Sun: Innovations to Harness Solar Energy and Power the Planet*. Sivaram has served as strategic advisor to the office of New York Governor Andrew M. Cuomo on Reforming the Energy Vision and was formerly a consultant at McKinsey & Company. Previously, he served as senior advisor for energy and water policy to the mayor of Los Angeles and oversaw the city's Department of Water and Power. Sivaram received degrees in engineering physics and international relations from Stanford University, and a PhD in condensed matter physics from Oxford University, where he developed third-generation solar photovoltaic coatings for building-integrated applications.

David G. Victor is a professor of international relations at the School of Global Policy and Strategy and director of the Laboratory on International Law and Regulation (ILAR) at the University of California, San Diego. He is also the co-chair of the Energy Security and Climate Initiative at the Brookings Institution. Victor cofounded and directs the ILAR research center, which works to understand how regulation, from climate change treaties to human rights accords, affects actors in international relations. Victor is a leading contributor to the Intergovernmental Panel on Climate Change, a UN-sanctioned international body, and is the chair of Southern California Edison's community engagement panel for decommissioning of the San Onofre Nuclear Generating Station. Previously, he served as director of Stanford University's Program on Energy and Sustainable Development at Stanford University, and was a professor at Stanford Law School and taught energy and environmental law. The author of *Global Warming Gridlock*, Victor received his BA from Harvard University and his PhD from the Massachusetts Institute of Technology.

·

www.ingramcontent.com/pod-product-compliance
Lightning Source LLC
Chambersburg PA
CBHW060502280326
41933CB00014B/2824

9 780876 097489